"As the last sight and sound of towns and people faded away, the trail narrowed, my own kind of trees and free wild places closed around me. The headlands stood black against the last sun-glow over the ocean, and the quiet mountains waiting to be climbed seemed to fold on and on forever...."

—Lillian Bos Ross
The Stranger: A Novel of Big Sur

CALIFORNIA COASTAL TRAILS

MEXICAN BORDER TO BIG SUR

JOHN McKINNEY

CAPRA PRESS
Santa Barbara, 1983

ACKNOWLEDGEMENTS

For their efforts in establishing the California Coastal Trail, as well as aiding the author in route-finding and fact-checking, I'm indebted to many members of the California Coastal Trails Foundation, particularly Susan Reed, Greg Gilbert, Jim and Helen McKinney. I'm particularly grateful to Bob McDermott, Foundation Trailmaster, for his advice and companionship on hundreds of miles of trail. Jim Blakley, naturalist and historian of the Los Padres backcountry, was a generous source of information on national forest trails. For her creative research, a tip of the cap goes to Judy Pritikin, and for proofreading above and beyond the call of duty, a thanks to Judy Young. Numerous officials of the California Coastal Commission and Coastal Conservancy offered guidance during the embryonic stages of the California Coastal Trail and I thank them. Another thanks to Kelty Pack. National Forest and State Park Rangers have, with the exception of a certain Malibu Creek State Park ranger who ticketed my car at an obscure trailhead on Christmas Eve, been universally helpful. —JOHN McKINNEY

CREDITS

Cover design and maps by Alex Marshall. Cover photography by Bob Werling. Photography by F.G. Hochberg on page 118. Drawings by Daniel Randolph on pages 8, 10, 11, 23, 48, 66, 121, 126, 155, 170, 199, 203. Drawings by Jane Howorth on page 176. Drawings from *California Coastal Access Guide* on pages 161, 165. Photographs from U.S. Forest Service on pages 16, 196, 197. Photographs by Cristine Argyrakis on pages 14, 25, 39, 42, 51, 54, 58, 60, 81, 111, 123, 142. All other photos by the author.

Typography by Jim Cook/Santa Barbara.

Copyright ©1983 by John McKinney
All rights reserved.
Printed in the United States of America.

Library of Congress Cataloging in Publication Data
McKinney, John, 1952-
   California coast trails.
   Contents: v. 1. Mexican border to Big Sur.
   1. Hiking—California—Guide-books. 2. Trails—
California—Guide-books. 3. Backpacking—California—
Guide-books. 4. Natural history—California.
   5. California—Description and travel—1981- —
Guide-books.
GV199.42.C2m35   1983   917.94   82-22129
ISBN 0-88496-198-2 (v. 1)

Published by
CAPRA PRESS
Post Office Box 2068
Santa Barbara, California 93120

CALIFORNIA COASTAL TRAILS
Volume I

I. INTRODUCTION
  Foreword 7
  Planning Your Coastal Hike 9
  Working for a Better Trail:
    The California Coastal Trails Foundation 11
  Your Legal Right to Hike the Coast 12
  Campgrounds/Accommodations 14
  Great Coast Hikers of the Past 15
  Equipping for a Coastal Hike 17
  Precautions on the Trail 20
  Precautions in the Surf 21
  Conservation of Marine Life 22
  How to Use this Book 23

II. SAN DIEGO COUNTY
  Highlights 25
  CCT at a Glance 26
  CCT: San Diego County 27
    DAY HIKES
  SD-1: Border Field Trail 39
  SD-2: Silver Strand Trail 40
  SD-3: Torrey Pines Beach Trail 41
  SD-4: Del Mar Beach Trail 43
  SD-5: Three Lagoons Trail 45
  SD-6: Bayside Trail 47
  SD-7: Sunset Cliffs Trail 48
  SD-8: North County Trail 49

III. ORANGE COUNTY
  Highlights 51
  CCT at a Glance 52
  CCT: Orange County 53
    DAY HIKES
  O-1: Aliso Beach Trail 63
  O-2: Crown of the Sea Trail 64
  O-3: Newport Trail 65
  O-4: Back Bay Trail 66
  O-5: Huntington Beach Trail 67
  O-6: Bolsa Chica Lagoon Loop Trail 68
  O-7: San Mateo Canyon Trail 70
  O-8: Bear Canyon Trail 72
  O-9: Bell Canyon Loop Trail 74

IV. LOS ANGELES COUNTY
  Highlights 77
  CCT at a Glance 78
  CCT: Los Angeles County 79
    DAY HIKES
  LA-1: Meet Catalina Trail 90
  LA-2: Cabrillo Beach Trail 92
  LA-3: Palos Verdes Peninsula Trail 94
  LA-4: Backbone Trail #1 (Will Rogers Park to Topanga Park) 96
  LA-5: Backbone Trail #2 (Tapia-to-Malibu Loop Trail) 98
  LA-6: Zuma-Dume Trail 100
  LA-7: Leo Carrillo Trail 102
  LA-8: Triathlon Trail 103

V. VENTURA COUNTY
  Highlights 105
  CCT at a Glance 106
  CCT: Ventura County 106
    DAY HIKES
  V-1: La Jolla Valley Trail 112
  V-2: McGrath Beach Trail 114
  V-3: Emma Wood Trail 115
  V-4: Anacapa Island Loop 116
  V-5: San Miguel Island Loop 118

VI. SANTA BARBARA COUNTY
  Highlights 121
  CCT at a Glance 122
  CCT: Santa Barbara County 122
    DAY HIKES
  SB-1: Summerland Trail 143
  SB-2: Santa Barbara County Beach Trail 145
  SB-3: Goleta Beach Trail 149
  SB-4: Seven Falls Trail 151
  SB-5: Jesusita Trail 152

SB-6: Gaviota Peak Trail 153
SB-7: Point Conception Trail 154
SB-8: Point Sal Trail 156

VII. SAN LUIS OBISPO COUNTY
Highlights 159
CCT at a Glance 160
CCT: San Luis Obispo County 160
DAY HIKES
SLO-1: Nipomo Dunes Trail 171
SLO-2: Coon Creek Trail Trail 173
SLO-3: Montana de Oro Bluffs Trail 174
SLO-4: Valencia Peak Trail 175
SLO-5: Morro Bay Sandspit Trail 176
SLO-6: Black Mountain Trail 178
SLO-7: Leffingwell Landing Trail 179

VIII. MONTEREY COUNTY (South)
Highlights 181
CCT at a Glance 182
CCT: Monterey County (South) 184
DAY HIKES
M-1: Salmon Creek Trail 194
M-2: Vicente Flat Trail 196
M-2: De Angulo Trail 197
M-2: Pfeiffer Beach Trail 199
M-2: Tanbark Trail 200
M-2: Cone Peak Lookout Trail 201
M-2: Prewitt LoopTrail 202
M-2: Mill Creek Trail 203

IX. APPENDIX
More Coastal Hiking 205

INTRODUCTION

Foreword

Other states have snowy peaks, wild rivers and vast deserts, but only California has a coastline of such length and diversity. This is a book about hiking that coastline from the Mexican border to Big Sur, an invitation to walk along land's end for six miles or six hundred, to experience what many consider to be the greatest meeting of land and sea in the world.

This book describes coastal trails—around precious wetlands, along precipitous bluffs, across white sand beaches, around bays, across islands, to lighthouses and through redwoods. It also introduces you to a new trail—the CALIFORNIA COASTAL TRAIL (CCT), designed to guide ambitious hikers from the Mexican border to the Oregon border along a 1400-mile system of interconnecting beach and coastal range trails. For the less ambitious, the trail will provide days, weekends, and weeks of exploration and recreation along one of the most unique environments on earth.

CCT visits nearly every natural attraction (and some unnatural ones) on the California coast. As the trail winds its way from Border Field State Park on the Mexican border to Pelican State Beach on the Oregon border, it passes through a hundred state beaches and parks, and a few hundred more refuges, reserves, county beaches, city beaches, and national seashores. CCT climbs inland through the chaparral-cloaked Santa Monica and Santa Ynez Mountains, Ventana and San Rafael Wilderness Areas in the Los Padres National Forest, fern-covered canyons in the Santa Cruz Mountains and majestic redwoods in Redwood National Park.

CCT not only provides access to natural attractions, it provides access to man-made attractions: Newport Bay, Marina Del Rey and Sausilito; the craziness of Pacific Beach and Venice; the tanning centers of Mission Beach, San Clemente, Santa Monica. Quaint fishing villages, clam festivals, Cannery Row, the art colonies of Laguna and Carmel; a cornucopia of lifestyles from Esalen to Eureka.

CCT hugs the coast and coastal slopes as closely as possible. Obstacles include military installations, two nuclear power plants, and places where sheer cliffs or residential and commercial structures prevent beach walking. CCT offers a "European style" hiking experience; campgrounds are often well developed, towns border pristine scenic areas, and usually little distance separates towns, rural areas and wilderness.

Lastly, and perhaps most importantly, this book is about the California coastline, its sand strands and cobble shores, its fists of mountains rising above the surf, its tides and tidepools, its past, its present, its future.

It's possible, in the course of a day, to hike over wildflower covered bluffs, along silver crescents of sand, detour around high-rise condominiums and walls of stucco, walk beneath barbed wire-covered sea walls, pass beneath the timbers of a historic wharf, picnic on a trackless pocket beach secreted between reefs, to watch the sun set over a rocky point covered with a colony of leopard seals catching the last rays of day. It's also possible to spend a day on the California Coastal Trail pondering these contradictory images and asking yourself some questions: Why is so much of the coastline owned by a few men of narrow vision and wide property interests? Shouldn't the coastline be a commons, protected and enjoyed by all citizens? Can't cows, condos, and hikers co-exist?

You might spend a day hiking the coast and thinking about these questions, pondering them until you're dizzy, until the sun drives all visions from your brain except that of a cold beer. You may feel at day's end that you've made no progress, but you have—10 or 15 mile's worth, in fact. You hiked past an estuary, a bay, and a beach, and some of the most giant coriopsis you've ever seen. By seeing the many "coasts" of California you begin to understand the whole. The more you hike the more various the habitats you encounter and the more you realize how interdependent they are. These habitats are links in a single chain, dependent on mankind for their continued survival.

What the California Coastal Trail provides is a macroview of the coastline, a slow motion study of land and sea, a link between ourselves and the edge of the continent.

May the chain remain unbroken.

Planning Your Coastal Hike

The approximately 600 miles of CCT described in this volume can be divided into three catagories: 35% of the trail follows undeveloped beach and bluffs, 30% travels urban/suburban or developed beach and bluffs, 35% ventures inland through coastal range wilderness. To help you plan your hike, a brief overview of the CCT follows.

CCT begins at Border Field State Park and visits San Diego Bay, Coronado Island, La Jolla, Torrey Pines, Cardiff and Carlsbad. There's a beach for everyone along San Diego County's 76 miles of shore and CCT visits all of them. There are beaches for swimming, surfing, sailing and diving, beaches for family picnics and beaches for romance, beaches for catching waves and beaches for catching fish.

From Oceanside, CCT skips past Camp Pendleton and resumes at San Onofre State Beach. From here, CCT visits San Clemente, Laguna Beach, Newport Harbor, Huntington Beach and several wide Orange County beaches. We pass the Long Beach Harbor where the Queen Mary is docked, visit the ancient lighthouse in San Pedro, round the rocky Palos Verdes headlands, and begin a slow saunter up the wide L.A. County beaches of Manhattan, Redondo and Hermosa. We hike past modern Marina Del Rey, the largest man-made pleasure craft harbor in the world, and the bustling Venice Beach boardwalk. From palm-lined Santa Monica Beach, CCT travels the secluded Malibu Coast and makes inland forays to sample the delights of the Santa Monica Mountains.

From Point Mugu, CCT skirts the Oxnard Marina and continues along a number of state beaches in Ventura County. Heading almost due west now, CCT passes through Carpinteria and arrives in Santa Barbara.

From Santa Barbara, CCT switchbacks inland into the Santa Ynez Mountains and begins a long traverse of the Los Padres National Forest. CCT visits Santa Cruz Station, Mission Pine Basin, Manzana Narrows and many other beautiful spots in the rugged San Rafael Wilderness.

Just north of Point Sal, CCT returns to the coast and heads north over vast sand dunes, forging on to the clams and clammers at Pismo Beach, then on to the hot springs at Avila Beach. Morro Bay, Cambria, San Simeon, and many other rocky and sandy central coast beaches are passed on a beautiful blufftop trail.

CCT climbs inland along the coastal slopes of the northern Los Padres National Forest. The trail leads through the Ventana Wilderness, climbing past cool redwood groves to summits that provide magnificent views of the Big Sur coastline. The first half of CCT concludes at Big Sur, the heart of California.

(Volume II of *California Coastal Trails* provides an account of CCT from Big Sur to the Oregon border.)

In the last few years, the California Coastal Commission has made tremendous progress in securing perpendicular access to the coast. Official accessways are posted with signs bearing the Commission's logo—a bare foot and a wave. Coast hikers can be grateful for the Commission's efforts in securing access and the laws requiring them, for without them, joining the California Coastal Trail would be virtually impossible.

Hikers can join the CCT from a thousand different places. Highway 1 and its side roads bring the motorist close to the coast. Greyhound bus service is excellent. Private and county bus lines extend to many trailheads in this volume. Amtrak has a San Diego-Los Angeles-San Francisco run with stops at Del Mar, Oceanside, San Clemente, San Juan Capistrano, Los Angeles, Oxnard, Santa Barbara, and San Luis Obispo.

CLIMATE: Unlike most other major hiking routes in America, the CCT is an all-year trail. Except for a portion of the trail passing through a section of the southern Los Padres National Forest that is closed during fire season, one can hike any part of the CCT at any time of the year. Of course the coast can be quite rainy in winter, and the farther north you hike the more rain you'll encounter. But winter, as many realize, is often the best of seasons to visit the coast. Minus tides make wonderful tidepooling, storms cast up all manner of marine treasure, and the big waves are a stirring sight.

The five Southern California coastal counties of San Diego, Orange, Los Angeles, Ventura, and Santa Barbara have one of the mildest climates in America. The coast from the Mexican border to Point Conception has a Mediterranean climate—rainy winters, and long hot summers tempered by sea breezes. Fog and low clouds are common and show a considerable degree of seasonal variation. It may be clear all day through two-thirds of the winter; however, in June and July, more than half of the mornings may be foggy. In winter, northern winds dominate, while summer is marked by gentle westerly winds, which typically increase in intensity in late afternoon. Water temperatures vary from a January low in the low 50s F to an August/September high of 75° F.

The southern Los Padres National Forest is characterized by hot summers and wet winters. During the rainy season the creeks and rivers rise to substantial proportions. Check with rangers about river crossings, particularly the Sisquoc. Sometimes snow dusts the upper peaks in winter. Summers are hot and dry and parts of the Los Padres are closed during fire season, approximately July 1 to first rain. Hiking in the southern Los Padres is best in spring and fall.

The central coast from Point Conception to Big Sur is cooler, wetter, and more foggy than the coastline to the south. The northern Los Padres receives a great deal of rain, as the presence of redwoods attest. Rains may begin as early as October, but the big storms usually hold off until December. Stream crossings during times of high water can be very difficult. Summer months are hot, except along the lower elevations in coastal canyons, where the summer fog cools the air. *Where CCT goes in the northern Los Padres, there's no fire closure.*

Working for a better trail: The California Coastal Trails Foundation

For decades hikers have dreamed of a continuous trail along the beaches, bluffs, and coast ranges of the California coast. Proposals for a "Pacific Coast Trail," an "Anza Historical Trail," and a "Coastal Trail," have surfaced before, only to become buried in bureaucracy. With this volume and the second, covering Big Sur to the Oregon border, the dream is set in type, a proposal begins evolving into a pathway.

The route I've devised, with the help of many dedicated hikers, is far from perfect. The California Coastal Trail, by necessity, is a compromise, because no sign has yet been erected, no land has been dedicated. What I've outlined is a possible route, something you can hike right now to get up the coast with a maximum amount of enjoyment of natural features and a minimum amount of hassle from private property holders. It's my hope and the hope of thousands of hikers, surfers, fishermen, and coast lovers of all sorts, that before too many years pass, an optimum route will be developed with the cooperation of state officials. This would be a signed and sanctioned trail, with rights to hike the coast, coastal bluffs, and coast range unequivocally established. Ideally, the trail would be supplemented by additional trail camps and hostels.

The optimum California Coastal Trail envisioned by hikers will follow a more inland route than the one described in this guide. Hikers will someday be able to leave Will Rogers Park in the Santa Monica Mountains, travel west across this range to Point Mugu, cross the Oxnard Plain, and enter the southern Los Padres National Forest in Ventura County. CCT will pass through the Santa Barbara backcountry, proceed northerly through the Santa Lucia Mountains and Wilderness and return to the coast just above Morro Bay. From here the trail will travel along the coast to the Monterey County line and re-enter the Los Padres National Forest. CCT will then stay on the coastal range ridges, traversing the Big Sur backcountry to Monterey.

Obviously, much work will be required to convince government agencies and private property holders to cooperate with each other—and with hikers. Working with these groups to establish an official CCT is the California Coastal Trails Foundation. For the latest trail update or for information on how you can get involved in establishing the CCT, send a self-addressed stamped envelope to:

> Trail Coordinator
> California Coastal Trails Foundation
> P.O. Box 20073
> Santa Barbara, CA 93120

Your Legal Right To Hike The Coast

The law-abiding hiker may question whether he or she will be breaking any laws by hiking shoreline stretches of the CCT.

The answer is a qualified NO.

The California Constitution says that no owner of "the frontage of tidal lands or a harbor, bay, inlet, estuary, or other navigable water in this state shall be permitted to exclude the right-of-way to such water whenever it is required for any public purpose...and the Legislature shall enact such laws as will give the most liberal construction to this provision so that access to the navigable waters of this state shall always be attainable for the people..."

The courts have repeatedly held that recreation, including hiking, is among the "public purposes" intended by the Constitution.

In 1972, the people of California, impatient with the Legislature's failure to enact protective coastal legislation, adopted by initiative Proposition 20

The Coastal Zone Conservation Act of 1972. It was most explicit about who has the right of access to the beach. The second paragraph of the act states: "The people of the state of California hereby find and declare that the California coastal zone is a distinct and valuable natural resource belonging to all people..." This section has stood up in court under the challenge of those who read this as an unconstitutional infringement on property rights.

Traditionally, the courts and legislatures have held that beach access should be open to all people. There is no unanimity on how this access should be secured, but the principle is understood by most in local and state government. The California Supreme Court has said that beaches become publicly available if they have been used by the public for at least five years.

Public access to and along the ocean shore is not a concept originated by the Coastal Commission and the court system however; it goes back to pre-European North America, continues through the Spanish era and is entrenched in the present California constitution and statutes. Coastal access advocates can argue that the first European settlers on our shores found the original aboriginal inhabitants using the foreshore for clam-digging and the dry sand areas for their cooking fires. Californians continued these customs after statehood. Thus from the time of earliest settlement to the present day, the general public has assumed both the wet and dry shoreline is part of a public beach, which we all hold in common.

On a less philosophic and more practical level it's safe to say that in California the ocean and the beach below the "mean high tide line" belong to the state and are always available for use by swimmers and hikers. The determination of the "mean high tideline" at a particular place is a job best left to oceanographers, lawyers and the Coastal Commission—it's a legal Pandora's box. Recent court decisions are extending the legally hike-able portion of the beach to include sand, rocks and bluffs up to the first line of terrestrial vegetation.

Access to the legally hike-able parts of the beach can be another story. Owners of private beach-front property can forbid trespassing either personally or with fences and signs and violation of their wishes is a misdemeanor.

Hiking difficulties arise when homes are built so close to the ocean on flat beaches that the tideline moves up to the house due to beach erosion or waves pound across the property during storms. This situation arises in Malibu and Oxnard where some homes have had to be placed on stilts because the ocean moved in.

Where private property stretches unbroken along several miles of beach, hikers are effectively prevented from reaching the shoreline except by walking below the mean high tide line from a public access point, the nearest of which may be miles away. The California Coastal Commission has made great strides in recent years in identifying existing access points and condemning strips of land to enable the public to reach the beach.

The point is that although the coast is there for you to hike, reaching trailheads is often a problem.

Wherever possible I've indicated where you come close to private property in walking the CCT. However, in some cases, information about land ownership is confusing. The author, publisher, and California Coastal Trails Foundation wish to make it clear that nothing contained in or implied in this guidebook is to be construed as encouragement to trespass; furthermore, the author, publisher and California Coastal Trails Foundation disclaim all responsibility for the actions of any person who uses the information contained herein to further any illegal activity whatsoever.

Campgrounds/Accommodations

One of the ways in which a CCT hiker needs to be flexible is in his or her choice of accommodations. Most coastal state parks have recently added special hiker-biker campgrounds but some have not, so you might find yourself in a developed campsite, tenting between two motorhomes. Also, there are a couple of stretches along the trail where it is extremely difficult to travel from one campsite to another in the course of a day. Los Angeles County, for example, has but one public coastal campground—Leo Carrillo Beach at the border with Ventura County. Other troublesome campless stretches include the Santa Maria River as CCT descends from the Los Padres to the sea, and Morro Bay to San Simeon.

Until more hostels and campgrounds are established, and until the official CCT is established further inland, a discussion of overnight accommodations is in order.

—You can take a motel room for a night. This option may strike some CCT hikers as ridiculous, contrary to the ethos of backpacking, etc., but it's an option many long-distance bicyclists choose when in urban areas and it might be your only choice. A night's lodging in a week of coast walking can expediate your trip and can be looked upon as either a treat or necessary evil, depending on your point of view.

—You can skip ahead to the next campground. Using alternate transportation, you shouldn't have much difficulty getting to an official campground by the end of the day.

—You can sleep on the beach; however, in some places it's illegal and others inadvisable. I can't condone sleeping on the beach, particularly in urban areas from San Diego to Ventura. Secluded beaches in Santa Barbara and San Luis Obisop are a better bet. If you must camp on the beach, camp well above the high tide line and observe traditional sanitary customs.

Hotel del Coronado

Great Coast Hikers of the Past

Coast travelers of years gone by can provide inspiration for those contemplating a trek along CCT. In 1776, Juan Bautista de Anza led an expedition from Mexico to Monterey, to scout the land for crown and church and determine how best to civilize Alta California.

The Anza Trail or Mission Trail is similar in route to the CCT in stretches from Ventura to Gaviota and from the Santa Maria River to Pismo Beach. Travel writer Frank Riley heroically attempted to follow the Anza Trail to commemorate the trail's bicentennial. He met a lot of obstacles, particularly along the Santa Barbara coast. "I know of no other historic trail on earth that has been so effectively closed to the public," says Riley. In 1976, the Anza Re-enactment Committee rode their horses along stretches of the Anza Trail.

Another inspiring coast traveler was Joseph Smeaton Chase, who in 1911 rode his horse Chino from the Mexican border to Oregon. His classic 1913 book, *California Coastal Trails* described his trip. Here is Chase along the central coast:

> "On leaving Morro I found myself entering that little-known stretch of mountain country which borders the Pacific closely for a distance of about a hundred miles. For most of that distance there are no roads and few settlers, while the trails are rough, steep, and often so little traveled as to be difficult to follow. Further, no maps of the region were to be had. Many persons had told me that I should never get through without a guide."

In 1970, Don Engdahl, a journalist from Santa Rosa California, backpacked most of the California Coast. Engdahl's stories, in the San Francisco *Chronicle,* about the gorgeous scenery and the barbed wire fences he found strung across bluffs and beaches, dramatized the need for public access to the coast. Near Fort Bragg, Engdahl was evicted from state-owned tideland by irate summer home dwellers and ranchers.

Carl and Harriet Ghormley and family finished walking the entire California coast in July 1982. They began in the summer of 1964 and walked a little bit of coast each summer. After 18 years and 177 days of walking, they reached Pelican Beach on the Oregon border!

Introduction

Equipping For a Coastal Hike

A discussion of hiking equipment is beyond the scope of this guide. There are many good books on the subject. However, coast and coast range hiking creates some special equipment needs:

BACKPACKS: It's the author's opinion that an internal frame pack with its backhugging qualities and narrow profile is the best choice for the sand-walking, rock-hopping and trail hiking required along the California Coastal Trail. A good one is infinitely adjustable. Compression straps keep the load from jiggling. If adjusted properly, the load will be stable, the weight distributed uniformly. Take a good look and try to test-hike before purchase. Pay particular attention to the suspension system. The load should be focused on your center of gravity, regardless of your walking motion.

The chief disadvantage of internal frame packs is that they tend to run smaller and have less carrying capacity than a conventional frame pack. However, except for CCT's long stretches through the San Rafael and Ventana Wilderness Areas, "civilization" is near the trail, so you'll rarely need a huge pack.

DAYPACKS: A good one will last a lifetime. Those bike bags or book bags of cotton or thin nylon won't hold up well. Shoulder pads and a hip band for support are nice features. Get one with a tough covered zipper. Double-O rings slip and aren't the greatest shoulder strap adjusters. Get tabler buckles; they don't slip and they adjust quickly.

FOOTWEAR: Leave those three-pound mountaineering boots at home. Investigate the new running shoe-hiking boot hybrids. They offer more support and stability than a running shoe, yet are amazingly light and well ventilated. They vary in style; some look like reinforced running shoes, others like hiking boots with fabric uppers. Look for a good sturdy toe box and let the strength of your ankles determine whether you require a boot cut high or low in the back.

A disadvantage of the new lightweights is that they're not waterproof, a consideration if you're walking wilderness portions of the CCT in winter. They are water repellent, however, and will get you through wet meadows and mudflats. If you purchase one of the low-cut models, expect to get your socks quite dirty and to stop to empty rocks, dirt, and sand out of your shoes.

Ecologically, the non-Vibram sole is much kinder to the trail, because it displaces less earth. A lightweight boot in most cases is kinder to your feet too and you'll find the weight saving will allow you to stuff a conventional pair of running shoes in your back to wear on those easy sand beach stretches.

CLOTHING: A T-shirt and a cotton shirt that buttons down will give you a lot of temperature-regulating possibilities. Add a wool shirt and a windbreaker with a hood and you'll be protected against the sudden changes of temperature that occur on the coast and in the coast ranges.
- Shorts are useful much of the year in Southern California.
- Swimsuits made of a quick-drying material are comfortable for beach walking.
- For cooler days or walking through the Los Padres brush, a sturdy pair of long pants is necessary.
- Hats prevent the brain from frying and protect from heat loss when it's cold.
- Sunglasses are particularly needed when walking over white sands or on hot, exposed slopes.
- Raingear—For the ultimate in protection and freedom of movement a gore-tex™ rain suit is a good investment, particularly if you plan on hiking the CCT during the rainy season. Other times of the year you can get by with a poncho—a cheap vinyl one is okay unless you walk through brush with them. The Boy Scout versions aren't too bad.

FIRST AID KIT: A standard kit supplemented with an ace bandage in the event of hiker's knee or sprained ankle. Take moleskin for blisters. Sun screen or tanning lotion will keep you from getting too red-faced and is absolutely essential on the California Coastal Trail. Insect repellent won't stop mosquitos from buzzing around but it will inhibit their biting.

WATER: It's still possible to drink from most Los Padres backcountry springs and streams without ill effect, but each individual water source should be carefully scrutinized. Many hikers assume water is pure and 48 hours later have a queasy feeling that tells them their assumption was wrong. Water may harbor the organism *Giardia Lamblia,* one of the causes of "traveler's diarrhea." When you approach a water source think about what may be upstream. A campground? Cows? Bring purification tablets and use them if you have the slightest doubt about water quality.

MAPS: The maps in this gude are sufficient for most purposes in Southern California. When traveling inland through the coast ranges, you are strongly urged to take a Los Padres National Forest map, available for a small fee at ranger stations. These general maps show rivers, roads, trails and camps and point out where trailheads are located and where travel is restricted during fire season. You may also wish to pack a few topographic maps when hiking in the coast ranges. These maps show terrain and elevation in great detail, which the maps in this book do not.

TIDE TABLES: There are two high tides and two low tides every 24 hours 50 minutes. (It would be convenient for hikers to make it an even 24 hours, but the tides are governed by the gravitational pull of the moon and sun over which we have no control.) The times of these tides are predicted and published each year by the U.S. Hydrographic Office. Local tide booklets are available free at many marine hardware stores, dive shops, and sporting goods stores. *Having the right tide table in your possession is crucial, for some parts of CCT are difficult, even impossible, at high tide.* As you hike the CCT, you must pick up a new booklet every 100 miles or so, for a tidebook issued for Mendocino has no more value in San Diego than a Utah road map. Last year's tide table is no more useful than last year's calendar.

The time to go hiking, tide pool exploring, clamming, or seashell collecting is at low tide. You'll want to plan your day so that you begin hiking a few hours before low tide and finish a few hours after.

Precautions On The Trail

POISON OAK: This infamous plant grows abundantly through the coastal ranges up to an elevation of 5,000 feet and is present on many ocean bluff trails. It may lurk under other shrubs or take the form of a vine and climb up a redwood or an oak. Poison oak's three-lobed leaves resemble leaves of the true oak. The leaves are one to four inches long and glossy, as if waxed.

All parts of the plant at all times of the year contain poisonous sap that can severely blister skin and mucous membranes. Its sap is most toxic during spring and summer. In fall, poison oak is particularly conspicuous, its leaves turning to flaming crimson or orange. However, its color change is more a response to heat and dryness than season; its "fall color" can occur anytime, particularly in Southern California. In winter, poison oak is naked, its stalks blending into the dull hue of the backcountry.

There are a multitude of remedies. A bath with one-half cup sea salt and one-half cup of kelp helps dry the oozing. A dip in the ocean can help too. A few tablespoons of baking soda added to a tub of lukewarm water soothes the itch as well. Mugwort is also an effective panacea. Its fresh juice applied directly to the pained area relieves itching. Then, of course, there's always calamine lotion or cortisone cream.

RATTLESNAKES: Despite the common fear of rattlers, few people see them and rarely is anyone bitten. The red diamond and southern pacific rattlesnakes are found in coastal regions visited by CCT.

Getting to a hospital emergency room is more important than any other first aid. Keep the site of the wound as immobilized as possible and relax. Cutting and suction treatments are now medically out of vogue and advised only as a last resort if you're absolutely sure you can't get to a hospital within four hours.

BEES: More fatalities occur from allergic reactions to insect stings than from rattlesnake bites. People allergic to bee stings can get a desensitization shot and a specially-equipped bee kit from an allergist.

TICKS: They're one quarter inch long and about the same color as the ground, so they're hard to see. Ticks are usually picked up by brushing against low vegetation. When hiking in a tick area it's best to sit on rocks rather than fallen logs. Check your skin and clothing occasionally. You and your companion can groom each other like monkeys at the zoo. If one is attached to the skin it should be lifted off with a gentle pull. Before bathing, look for ticks on the body, particularly in the hair and pubic region.

Precautions In The Surf

California waters contain very little hazardous marine life. Dangerous sharks are extremely rare near shore, although harmless varieties are not uncommon.

STINGRAYS may be encountered, particularly in late summer. They lie about the bottom in the surf zone and if stepped on may inflict a painful wound. If stung, cleanse the area thoroughly to avoid infection.

SEA URCHINS appear in rocky tidal zones. Armored with brittle purple spines they grow 2 to 6 inches in diameter. If you step on an urchin the spines break off. You must remove the mildly poisonous spines with great care because if allowed to remain under the skin, they'll make the wound hurt for a long time. Dissolve the spines, which are made of calcium carbonate, in a weak acid such as vinegar, lemon juice, or, uh...uric acid. The latter suggests possible first aid procedure on a deserted beach.

JELLYFISH are common in the summer months. Avoid them. Their clear blue or purple umbrella-shaped floats are less than a foot in diameter, but their stinging tentacles may dangle ten or fifteen feet beneath the surface.... Traditional first aid is to rub the affected area with wet sand, wash it with ammonia, and apply burn ointment.

Conservation of Marine Life

With increased interest in marine biology and with so many people visiting California's coast, it's of vital importance to conserve and preserve the marine life inhabiting our rocky shores, sand beaches, and bays. All too frequently, collections of larger marine animals such as star fish, sea urchins and crabs end up in trash barrels after they begin to smell. Better to have let the animal live its life. Collecting for all but the most scientific purposes should be discouraged, and it is usually illegal.

Some general rules should be remembered whenever one is observing marine organisms. Rocks, which have been turned over, should be replaced in their original position, otherwise the plants and animals which were originally on the upper surface are now on the bottom and will die; the same, in reverse, holds for animals which were originally on the bottom of the rock. Whenever digging in the sand or mud for clams or other creatures, the material should be shoveled back into the hole, otherwise many of these organisms will also die because their habitat was disturbed. Remember each species has its own specific habitat and whenever we disturb its environment, chances are the organism will die.

Marine Reserves protected by the California Department of Fish and Game are posted with large green signs both at entrances to the Reserve and at different places along the beach. They state that only certain fish may be caught or that no fishing is permitted. Sometimes there are warnings not to remove rocks, shells, or other material.

How to Use This Book

California's coastal trails are organized county by county in this guide book. An overview of individual counties is offered at the beginning of each chapter. Included are summaries of natural highlights, terrain, transportation, accommodations and obstructions. A step-by-step account of the California Coastal Trail (CCT) follows. At the end of each chapter you'll find descriptions of day hikes and overnight hikes. Those hikes using the CCT are described with a minimum of detail and you are asked to refer to the preceding CCT description for a full account. Those hikes not on the CCT are described in more detail.

Hikers can use this book in other ways. Pick that section of coastline you've always wanted to explore and find a day hike to introduce you to the area. If you have a specific goal in mind, consult the table of contents for a description of the locale. Further hiking opportunities are listed in the appendix, county by county.

The California coast is often described as one of the most beautiful in the world. It is there for us to enjoy, if we use it wisely.

San Diego County

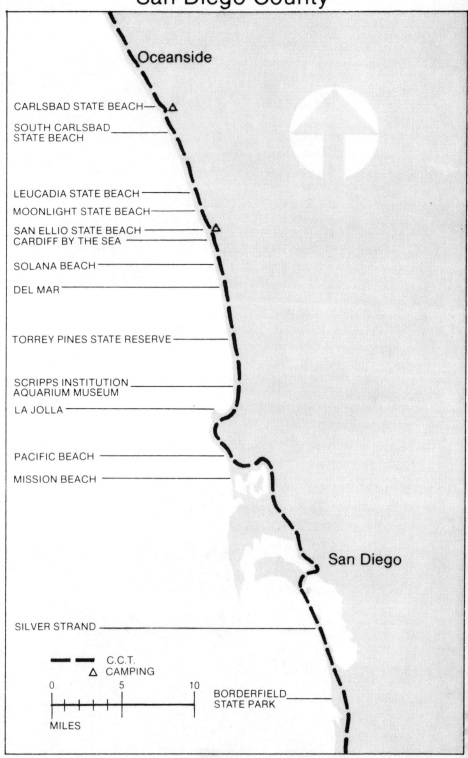

San Diego County

Cabrillo Lighthouse

It is said that California began in San Diego. In 1542, Portuguese navigator Juan Rodriguez Cabrillo landed on Point Loma and named the area San Miguel. When explorer Vizcaino arrived at the same spot in 1602, he re-named it San Diego.

San Diego County's shoreline extend approximately 76 miles upcoast from the Mexican border to San Mateo Point near the city of San Clemente. The county has an additional 27 miles of shoreline on Mission Bay. Each beach seems to have its own character—best surfing, clearest water, panoramic view, most birdlife, etc. The air and water temperatures are Mediterranean, the place names Spanish. It's easy to see why the county is an attraction for visitors who come for the sun, sailing, and historical romance.

The beaches near the border—Border Field, Silver Strand and Coronado —are wide sandy stretches on the west side of San Diego Bay. North of Mission Bay, the largest aquatic playground in the world, are more wide, tanning beaches, Mission Beach and Pacific Beach.

From Point Loma to La Jolla, rocks and reefs carve the oceanfront into a series of pocket beaches, dramatically altering waves and currents so that a few feet of movement can transport a hiker from a safe swimming area to a paradise for surfers to a frothy wave-swept cauldron. Some of the northern La Jolla beaches are sandy, but rock predominates further south. One of the world's premier oceanography centers, Scripp's Institute, is located in La Jolla.

North County is a shoreline of sandstone bluffs broken up by scattered lagoons. On the bluffs south of Del Mar grow the rare Torrey pines. Many of Southern Caliifornia's valuable tidelands are here: Penasquitos Marsh,

Buena Vista, Aqua Hedionda, Batequitos. Lagoons provide nesting and resting places for dozens of species of native and migratory waterfowl.

For the hiker, San Diego County offers miles and miles of easy beach-walking. Those wishing to pause along the CCT, may explore diving reefs, choose from a multitude of great body-surfing spots, go sailing, or simply doze and tan on the quiet beaches.

CCT at a glance

TERRAIN: Sandy Beach from border to Coronado. Rocky cliffs at Sunset Cliffs and La Jolla. North County a particular delight—sandstone cliffs, lots of lagoons. CCT stays right on beach.

OBSTRUCTIONS: No pedestrians are allowed on the Coronado Bay Bridge. You must take the bus. No trespassing is permitted on Camp Pendleton Marine Base. Take bus or train north from Oceanside. Some minor detours required to get past residential areas in La Jolla.

TRANSPORTATION: Amtrak stops in San Diego, Del Mar, and Oceanside. For bus transit information in areas south of Del Mar, call San Diego Transit at (714) 233-3004. For transit information from Oceanside to Del Mar, call the North County Transit District at (714) 484-2550 in Del Mar and Solana Beach and (714) 438-2550 in Oceanside and Carlsbad.

CAMPGROUNDS/ACCOMMODATIONS:
San Onofre State Beach
South Carlsbad State Beach
San Flijo State Beach
Fiesta Island Youth Camp (By permit only)
Point Loma Hostel
Armed Services YMCA Hostel (San Diego)
Imperial Beach Hostel

CCT: SAN DIEGO COUNTY

The California Coastal Trail (CCT) begins at the very southwest corner of America, at the monument marking the border between Mexico and California. When California became a territory at the end of the Mexican-American War of 1848, an international border became a necessity. American and Mexican survey crews determined the boundary and the monument of Italian marble was placed to mark the original survey site in 1851. Today the monument stands in the shadow of the Tijuana Bull Ring and still delineates the border between the United States and Estados Unidos Mexicanos.

During World War II the Navy used Border Field State Park as an air field. Combat pilots received gunnery training, learning to hit steam-driven targets that raced over the dunes on rails called Rabbit Tracks. Despite multifarious real estate schemers, the Navy retained control of Border Field until he land was given to the state in the early '70s.

Before you walk down the bluffs to the beach, take in the panoramic view: the Otay Mountains and San Miguel Mountains to the east, Mexico's Coronado Islands out to sea, and to the north—the Tijuana River floodplain, the Silver Strand, Coronado.

Hike down the short bluff trail to the beach, which is under strict twenty-four-hour surveillance by the U.S. Border Patrol. The beach is usually deserted, quite a contrast to crowded Tijuana Beach a few hundred yards south. As you walk north on Border Field State Park's mile and a half of beach, you'll pass sand dunes anchored by salt grass, pickleweed and sand verbena. On the other side of the dunes is the Tijuana River Estuary, an essential breeding ground, feeding and nesting spot for over 170 species of native and migratory birds. This is the first wetlands encountered along the CCT and one of the most pristine. At Border Field, the salt marsh is relatively unspoiled, unlike many wetlands encountered farther north, which have been drained, filled, or used as dumps.

Take time to explore this beautiful marsh. You may spot hawks, pelicans, terns and ducks, as well as many other nesting and shore birds. Fishing is good for perch, corbina and halibut both in the surf along Border Field Beach and in the estuary.

A mile and a half from the border you'll reach the mouth of the Tijuana River. Only after large storms is the Tijuana River the large swath pictured on some maps. Most of the time it's easily fordable at low tide.

Continue north along wide, sandy Imperial Beach, past some houses and short bluffs. Don Santiago Arguello once owned all the land from the tip of San Diego Bay to the Mexican border in the days before California became part of the United States. Imperial Beach was named by the South San

Diego Investment Company, in order to lure Imperial Valley residents to build summer cottages on the beach. Waterfront lots could be purchased for $25 down, $25 monthly and developers promised the balmy climate would "cure rheumatic proclivities, catarrhal trouble, lesions of the lungs," and a wide assortment of other ailments.

In more recent times, what was once a narrow beach protected by a seawall, has been widened considerably by sand dredged from San Diego Bay. There's good swimming and surfing along Imperial Beach and the waves can get huge. CCT passes Imperial Beach Pier, built in 1912 and the oldest in the county.

Six blocks north of the pier is Palm Avenue, where you'll find the recently opened Imperial Beach Hostel, the first of hostels available to northbound CCT hikers. Turn east a short ways to 170 Palm Avenue, Imperial Beach, CA 93032, (714) 423-8039.

North of Imperial Beach hikers reach the Silver Strand, whose sparkling sands actually belong to Border Field Beach. Sand is carried away from the border beaches by a longshore current which runs parallel to the coast. Along most of the California coast, this current is flowing south, but here it's flowing north because of eddies set up in the water by Point Loma. This unique circulation pattern formed Silver Strand, which separates San Diego Bay from the ocean.

CCT enters Silver Strand State Park, which administers the southern third of the narrow spit of land linking Coronado with Imperial Beach. It's an ultra-developed park, complete with shaded picnic tables, 400 fire rings and 2,000 parking spaces! Be sure to take one of the pedestrian underpasses beneath Highway 75 for a look at San Diego Bay. There's excellent swimming on the bay side, but keep an eye out for jellyfish.

Besides all this civilization, the state park also hosts such water fowl as Brandt's cormorants, gulls, terns, sanderlings and loons. California sea lions are numerous offshore and an occasional school of porpoises visits the area.

North of the state park the tide tosses up a lot of junk and beachcombing artists fashion it into driftwood and plastic sculptures. It's also a good beach to look for shells. CCT passes some "Star Wars"-style hardware belonging to the U.S. Naval Communications Station. At the north end of the Silver Strand, you'll first encounter man's architecture at its worst—the cell-block style of the Coronado Shores condominium development, then architecture at its best, the wondrous Victorian-era Hotel Del Coronado.

Charming Coronado has attracted beach lovers since the town was developed in the late 1800s. As in earlier days, visitors flock to the "Del" and the sands nearby, yet the beach is surprisingly uncrowded.

Before the San Diego-Coronado Bay Bridge was opened in 1969, chugging ferries carried passengers and cars across the bay, a tradition many regretted to see end. No pedestrians are allowed on the bridge, so you must board an American Sightseeing Bus (a private carrier) on their 901 Strand Route line at 4th and Orange Streets, Coronado. The bus crosses the bridge and makes its first mainland stop at Crosby and National Boulevard. Take Crosby west three blocks to Harbor Boulevard, following Harbor one mile along a bike lane. Hikers pass over a small bridge with not-very-inspiring sights of freight yards on the right and shipyards on the left.

Cross Harbor Boulevard at Embarcadero Marina Park. You'll pass grassy picnic areas and a fishing pier. Endure congested Seaport Shopping Center and continue along the waterfront. CCT passes the G Street Fishing Pier, then the Broadway Pier, which offers fine views of activity on the bay. The U.S. Custom's office is located at the end of the pier. Sometimes Navy ships at berth are available for public tours.

At 500 W. Broadway Street, five blocks east of the Broadway Pier on Harbor Drive, is the Armed Services YMCA Hostel. It offers ten private rooms and dorm lodging. Open all year around, (714) 232-1133.

CCT continues along the bay boardwalk past the Maritime Museum, home to three historic ships, the most famous of which is the Star of India. This three-masted ship is the world's oldest merchant ship afloat, launched in 1863.

Keep following the Embarcadero, a walkway providing views of the bay and Coronado. Jets from nearby Lindbergh field shiver the timbers of the walkway. At Laurel Street join the bikepath and follow it a mile to the turnoff to Harbor Island, a boating center with various marinas. If you'd like to explore, a walkway extends the length of the island, offering fine bay views and fishing off the rocks. CCT continues under a highway cloverleaf to Spanish Landing Park, a grassy area popular with picnickers. CCT follows the walkway along the sea wall, which offers fine views of the bay.

Cross the bridge at Harbor Drive over to Point Loma. Here there is a V-Intersection.

To reach the Point Loma Hostel: Bear right on Nimitz Blvd. Walk a mile on Nimitz. As you crest a hill, just past Chatsworth Street, you'll see the military green hostel on your right. Follow the sidewalk down to the hostel, located at 3790 Udall St. The hostel accommodates 26 men and 20 women; guests are provided with a room with bunk beds for two to eight people. Open at 4:30 p.m. with a midnight curfew. Contact: Point Loma Hostel, 3790 Udall St. San Diego, CA 92107, (714) 223-4778.

To continue on CCT: Bear left at V-intersection on Harbor Boulevard, then north (left) on Scott Street, passing the Fleet Anti-Sub Warfare School and the Commercial Basin, home to over 800 commercial fishing vessels. Continue along the Municipal Sport fishing pier to the turnoff to Shelter Island. The island's yacht basin holds 2,000 craft. The island's attractions include a nice sandy beach, picnic areas and a fishing pier.

CCT bears right (west) on Talbot Street for one mile across Point Loma to Catalina Boulevard. Go right one block, then make a left on Hill Street. Walk up (Cardiac) Hill then down to the "Rockpile" at Sunset Cliffs Park. CCT heads north atop the crumbly cliffs overlooking tiny pocket beaches between promontories of sculpted sandstone. Surfer trails lead down to rocky tidepools and sandy coves. The sunsets off these golden cliffs are superb.

Depending on the tide, you may be able to squeeze along between water and cliff face or you may have to detour inland for about 1/4 mile to Pescadero Beach, just south of Ocean Beach City Beach. At the long T-shaped Ocean Beach Fishing Pier, the beach begins to widen and become sandy. There's good body- and board-surfing here, particularly north of the pier. Clustered around the beach gym equipment is often an interesting sampling of exhibitionist Ocean Beach life.

Ocean Beach has long been a popular seaside community. To sell lots here in the 1880s, a couple of real estate promoters proclaimed in their ads: "The locality of Eden was lost to the world until Carlson and Higgins discovered Ocean Beach." The community was one of the few beach boom towns to survive.

CCT continues along Ocean Beach to the San Diego River Flood Channel. The mouth of the river, where rocks protrude into the water, is reserved for surfers. Sometimes the river mouth is filled in with sand and you can walk across, but usually you'll have to follow the river bank inland to the Nimitz Boulevard bridge and cross the river on the bikepath.

CCT winds around the Quivara Boat Basin. Improvise a route north then west using the red cement walkways around the basin to the Island Sport Fishing Center. You'll pass a whimsical sign insisting, *No dogs, bicycles, or high heel shoes.*

From here you could make a trip over to the southeastern shore of Mission Bay to Sea World. At this 75-acre theme park you can view aquarium exhibits and marine shows.

Take some time to explore Mission Bay, which compromises the largest aquatic park on the west coast. There's jet skiing, fishing and quiet bay beaches.

On the Portola Exhibition of 1769, Father Crespi referred to the bay as a "closed port" of San Diego harbor. It was later called False Bay until the federal government changed the name to Mission Bay in 1915.

CCT continues over Mission Bay Channel on the Glenn A. Rick Bridge, follows Mission Bay Drive for a short half mile and arrives oceanside at Mission Beach.

Mission Beach is San Diego's answer to the Jersey Shore. In front of the now defunct Belmont Amusement Center the boardwalk boogies with roller skaters, bicyclists, surfers, body-builders, and more healthfood, seafood and junkfood eateries and raft-renting establishments than anyone could wish for.

CCT leaves behind zany Mission Beach and reaches southern Pacific Beach. Winter storms shift the sand here so the waves seem to break twice, once at the end of Crystal Pier and again closer to the beach. Good body surfing here. Swimming is best south of the pier, which is a private fishing pier with a motel perched on top. Just south of the pier is Pacific Beach Park, a picnic area overlooking the ocean.

The north end of the Pacific Beach is quiet, less popular than the south because of the residential areas behind on the low bluffs. The beach narrows and grows rocky as you approach Tourmaline surfing park. Designated for surfers only.

It's a nice bit of rockhopping to continue north from Tourmaline. The low tides here are very low and the tidepool watching is superb. CCT rounds False Point and after a mile of bouldering reaches Bird Rock, a large guano-covered promontory. If the tide is exceptionally low, you may continue another half mile north to Sun Gold Point, which borders a small cove. Most likely you'll have to walk an inland route to La Jolla Strand Park, a seasonally sandy surfing beach, north of the point.

A few hundred yards north is Winansea Beach, which some say has the best surfing on the west coast. The surfing reserve is at the north end of the beach, where breakers sprint toward shore at heights of six feet on most days and get bigger on occasion. World class surfers train here. The shorebreak, on the right day, can rival the Wedge for tough and dangerous body surfing. Rock formations create shallow coves, ideal picnic spots for hikers.

Winansea Beach was made famous in Tom Wolfe's, *The Pump House Gang*. Hot Curl's statue was enshrined here. The famed pump house exists today, still pressurizing the county sewers.

> Simmons was a fantastic surfer. He was fantastic even though he had a bad leg. He rode out really big waves. One day he got wiped out at Windansea. When a big wave overtakes a surfer, it drives him right to the bottom. The board came in but he never came up and they never found his body. Very mysterioso.
>
> —Tom Wolfe, *The Pump House Gang*

Beyond Winansea, CCT picks its way over a series of rocky points, Seal Rock, Pt. La Jolla, Goldfish Point. At Seal Rock, CCT passes Children's Pool Beach, described in 1930s guidebooks as a safe place for children to swim; it's no longer quite so safe. The curving seawall was built with an arch at the south end. The ocean flowed through the arch, shoveling sand onto a sheltered beach between seawall and bluffs. Unfortunately, the wall began crumbling and the city filled in the archway, causing the sand to shift and altering the current so it now sweeps rapidly past the end of the seawall. Still, it's a nice beach and offers good swimming on most days—but watch for flying children.

Tucked away on the east side of Pt. La Jolla, is La Jolla Cove, a highlight of the long underwater park protecting the wondrous reefs located about twenty yards offshore. On a calm day, strong swimmers and snorkelers can explore the cove northwest to Goldfish Point, named for the brilliant and plentiful orange Garibaldi (which look like giant goldfish). La Jolla Cove is filled with nautilus, varieties of kelp, lobster, moray eels, and sea stars. All marine life is protected by the San Diego-La Jolla Marine Park, which extends from here to Torrey Pines.

North of Goldfish Point, CCT follows a dirt path called Coast Walk atop the La Jolla Bluffs. Hikers see panoramic views of the ocean, beach and ocean caves along the shore. Coast Walk takes you over several wooden pedestrian bridges back to Torrey Pines Road. Make the first left on Princess St., then immediately go right on Spindrift, which returns you to the ocean at La Jolla Shores Beach. From this point north to Oceanside, CCT makes no more forays inland; it's all beach walking.

La Jolla Shores Beach is a wide expanse of white sand. The water deepens gradually and offers good swimming. Surfers use the beach for two miles north to Scripps Pier. The La Jolla Indians once had a campsite here and ancient artifacts have been recovered from the shallow waters offshore.

At Scripps Pier, hikers may wish to take a break and spend some time at Scripps Institute of Oceanography located on the bluffs above. The Aquarium Museum is open 9:00 to 5:00 daily. Be sure to check out the dry land tidepool and see if you can identify fish passing by underwater video cameras.

CCT continues north of Scripps Pier on rocky beach, which soon widens and grows more sandy. Curry-colored cliffs rise higher and higher above you as you hike north. Atop the cliffs is a glider port, launching manned fixed wing gliders. Hang gliders also ride the thermals above the bluffs.

Below the handsome cliffs is Black's Beach, which for a short period in the mid-'70s was the first and only nude public beach in the country. Legally, suits are no longer optional, but most opt for none.

Two miles north of Black's Beach, you'll see a distinct rock outcropping, named, appropriately enough, Flat Rock. Legend has it this gouged-out rock, also known as Bathtub Rock, was the luckless site of a Scottish miner's search for coal. Today it serves as residence for many forms of tidepool life.

Just north of Flat Rock, a trail ascends the bluffs to Torrey Pines State

Reserve. The Reserve protects the rare Torrey pine, which occupies the bold headlands atop the cliffs. Several nature trails through the Reserve give the hiker an opportunity to view the pines and other plant life and afford great views of the coastline.

The *pinus Torreyana* clinging to the cliffs in strangely twisted and windblown shapes grow only here and on Santa Rosa Island. In 1850, Joseph Le Conte recognized the pine as a new species and contacted C.C. Parry, who was then busy with the Mexican-American boundary survey. The two men named the trees after a former professor at Columbia University, Dr. John Torrey.

A mile north of Flat Rock is Los Penasquitos Lagoon. Snowy plover, white tailed kite, light-footed clapper rail and the least tern are a few of the many species of waterfowl that can be observed here.

CCT continues north, following the sometimes wide, sometimes narrow beach over sparkling sand and soft green limestone rock. Holes in the limestone give evidence of marine life that once made its home there. Two miles past the lagoon you'll reach Del Mar Train Station.

> ...to the man or woman of leisure and means, a real resting place by the sea, free from noise, confusion, and ugly cheap details of the average beach resort. Del Mar is planned for the exclusive.
> —1912 sales brochure, South Coast Land Company

CCT travels over Del Mar City Beach, where "the surf meets the turf," as modern day advertisements boast. The long beach's northern end is directly across from the Del Mar Race Track and County Fairgrounds. It's legal to exercise horses here north to Solana Beach. You can watch trainers gallop their thoroughbreds if you forgot yours. Swimming and surfing are also popular.

CCT passes beneath houses perched surrealistically atop the handsome, but collapsing sandstone bluffs. If the tide is high, you may have to hike over rocks piled at the base of the cliffs to slow erosion. The beach is a favorite of surf fishermen and grunion hunters.

Beyond Del Mar is Solana (Spanish for "sunny place") Beach. One early settler claimed he was able "to regularly shoot rabbits, coyotes, and rattlesnakes" from his front porch. Today Solana Beach's serenity is threatened by condo development. At Solana Beach County Park there's picnicking and a life guard headquarters.

Half mile north of Solana, CCT traverses tiny Tide Beach County Park, which soon widens into Cardiff State Beach. At the south end of Cardiff are tidepools to explore. Across Pacific Coast Highway is San Elijo Lagoon. The Portola Expedition camped here on July 16, 1769.

Developed in 1912, Cardiff-by-the-Sea was no doubt named by an Anglophile, who borrowed the name from the seaport in Wales. British place names are on most of the town's streets. Originally the township was named San Elijo after the nearby lagoon. High-grade clay deposits were dug here and shipped out of state.

A bit beyond Cardiff is San Elijo State Beach, the southern-most campground in the state beach park system. The campground is located along the blufftop overlooking the beach. You'll find no hike-bike campground here but there are 171 campsites with tables, stoves, showers, the works. To reach the campground take the stairway from the beach to the blufftop.

Short jetties tame the waves at the state beach. At the lagoon, a long low point in the cliffs, the surf can be quite active.

As CCT heads up-coast, the bluffs rise higher and higher. Immediately south of the town of Encinitas, CCT reaches Sea Cliff County Park, a small bluff top park with picnicking. "Pipes" is an area just off the drainage pipes protruding from the southern cliffs. The offshore reefs here contain a bounty of sea creatures including kelp, bass, corbina, halibut, and abalone. The diving and surf fishing is excellent on a flat day. There are good lobster holes on the outer edge of the mudstone reef.

"Swamis" of surfing and Beach Boys fame is the reef area north of the stairs. It's named for the Lotus-shaped Self Realization Fellowship on the bluffs. Unconfirmed rumor has it that lifeguards sometimes play soothing music over the P.A. system to the hoards of surfers in the water. In short, the area around Sea Cliff County Park is one of the best surfing and diving areas in North County.

Beyond Swamis is the town of Encinitas, "the place of little oaks." When the Portola expedition marched northward in search of Monterey, it entered a valley of small oak trees two days after leaving San Diego and Father Crespi thus named it Canada de los Encinos.

The first Encinitas settler was eccentric Chicagoan Nathan Easton, who arrived in 1875. He kept bees, seemed to prefer a hermit's existence and built the first house in town—using old lumber and tin cans for shingles. He is credited with introducing the notorious Australian salt-bush to Southern California. He saw the bush advertised and sent to Australia for seeds. He

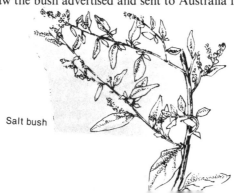
Salt bush

kept his pockets full and scattered them as he roamed the county with his mule team. He hoped the bush would make good forage for cattle, but it has only proved to be a nuisance wherever people have tried to cultivate the land.

Past the small town of Encinitas, the cliffs open up and fill with sand, forming wide sandy Moonlight State Beach. Body surfing is sensational here if you catch the waves at the right time of day. As you hike north of Moonlight, the sandstone cliffs resume and you soon cross the tiny beach at Seaside County Park.

On the bluffs above is the town of Leucadia, founded during the boom of the 1880s and named after a Greek island renowned for its beauty, fine wine and olives.

If you walk around town, you'll notice Greek and Roman street names—Hygeia, Eolus, Hermes, Vulcan. Leucadia and its streets were named by a group of British spiritualists, with a penchant for classical culture, who came to the U.S. seeking religious freedom. Leucadia Beach is wide and sandy, again the domain of the surfer. Half a mile north is Batequitos Lagoon, a refuge for endangered aquatic birds. La Costa is the township near the Lagoon.

Another mile north brings you to South Carlsbad State Beach, where 226 campsites with all the amenities are located atop the bluffs. The campground can be reached via stairways from the beach.

Two miles north of South Carlsbad is Carlsbad State Beach. As you approach, you'll spot what appears to be a giant lighthouse, but on closer inspection turns out to be a San Diego Gas and Electric power plant located at Aqua Hedionda (Stinking Water) Lagoon. The no longer odiferous waters of the lagoon make it a popular spot for boating and swimming. Fishermen try their luck at the Encina Fishing Area along the mouth of the lagoon.

CCT pushes onward past the state beach over sandy and rocky shore lined with bluffs. You pass Carlsbad City Beach. The resort town of Carlsbad, which in the nineteenth century attracted visitors to its healing sulphur springs, is today the largest of the coastal towns lining old Highway 101.

At the north end of Carlsbad, you encounter yet another lagoon, Buena Vista, a natural preserve and bird sanctuary where fishing is permitted. The wetland can be viewed from several points along its perimeter. As CCT arrives at south Oceanside Beach, you can see tiny La Salina Lagoon, where a variety of bird species can be observed from a short path.

According to local tradition, Oceanside was a community that just named itself; prior to 1884 it was family custom "to go oceanside" here on outings. Oceanside City Beach, populated with Marines on R and R

Oceanside Pier

features a 1900 foot long municipal pier, lighted at night for fishing and surfing. Once past the pier, Oceanside Harbor comes into view and one can see the boats entering and leaving the boat basin.

When CCT reaches the jetty south of Oceanside Harbor at the mouth of the San Luis Rey River, the hiker must make a decision on how best to conduct *THE CAMP PENDLETON DODGE*. A few options follow:

1) From Oceanside Beach, walk to the Amtrak station at 117 N. Cleveland (722-4622). Board the northbound train, which runs several times a day, and get off at the first stop, San Clemente. The train stops right at San Clemente pier.
2) From Oceanside Beach, walk to the Greyhound Bus Station at 217 N. Tremont (722-1587). Buses run frequently. Resume the CCT in San Clemente.
3) Hitchhike north on Interstate 5 to the second exit north of Oceanside, the Las Flores Viewpoint. Hike north along old Highway 101 (closed to traffic) two miles to the break-away-gate and another 1¾ miles beyond that to San Onofre State Beach; you'll be following the evacuation route planned for campers at San Onofre State Beach in the event the San Onofre Nuclear Generating Plant has a melt-down. At present this route is technically illegal for pedestrians, although park rangers at San Onofre say it will soon be open for public use. When you reach another gate at the boundary of the state park, a path with the less-than-inspiring name of Trail 6 branches off to the west and takes you back down to the beach.
4) Hitchhike to San Onofre State Beach north on I-5 to the Basilone exit. Resume the CCT from the north end of San Onofre Beach. If you want to camp at San Onofre, you'll have to hike back south a mile.

Camp Pendleton, in short, is the single largest impediment to the CCT in Southern California. Some imagination and patience are required to deal with it. A twenty-mile bike route across the marine base is available for bicyclists, but nothing for hikers as yet.

San Onofre is a place of steep bluffs, overlooking a wide sandy beach and magnificent surf. Unfortunately, the peaceful ambience of this beach has been completely annhilated by the giant nuclear spheres of the San Onofre Nuclear Power plant situated on the beach and bisecting the state park. It can be unsettling when park rangers give you a map with evacuation instructions—just in case the reactor melts down.

San Onofre State Beach South has a large campground along Old Highway 101. The hike-in campground at Echo Arch is reached by a short loop trail just west of the park entrance kiosk. The trail camp is located on a terrace between the beach and the bluffs. It's a little too close to the Nuke for my tastes.

CCT continues up-beach from the campground and soon deadends at the power plant. A seawall, constructed to prevent the reactor from being washed away, blocks beach access at the power plant. Rumor has it that a lighted catwalk will be constructed passing through the plant and connecting northern and southern portions of the state beach.

After detouring around the power plant you will rejoin San Onofre State Beach north, a wide sandy beach popular with surfers, clammers and surf fishermen. A private campground is located on the bluffs of this beach. At the north end of the state beach is San Mateo Creek, easily fordable under most conditions. The creek mouth is a gathering place for migratory birds and surfers.

Here is famous Trestles Beach, one of the finest surfing areas on the west coast. When the surf is up, the waves peel rapidly across San Mateo Point, creating a great ride. Before the area became part of the state beach, it was restricted government property belonging to Camp Pendleton. For well over twenty-five years, surfers carried on guerrilla warfare with U.S. Marines. Trespassing surfers were chased, arrested and fined, and on many occasions had their boards confiscated and broken in two. Find a veteran surfer and he'll tell you dramatic tales from the '50s, '60s and early '70s about escapes from jeep patrols and guard dogs. Many times, however, the cool jar heads would charitably give surfers rides while out on maneuvers.

San Mateo Point is the northernmost boundary of San Diego County, the beginning of Orange County. When the original counties of Los Angeles and San Diego were set up in 1850, the line that separated them began on the coast at San Mateo Point. When Orange County was formed from southern Los Angeles County in 1889, San Mateo Point was established as the southern point of the new county.

San Diego County

Border Field Monument

Border Field Trail (SD-1)

Border Field State Park to Tijuana River, 3 miles RT
Border Field State Park to Imperial Beach, 6 miles RT

Much of the Tijuana River Estuary, one of the few unspoiled salt marshes left in Southern California, is within Border Field's boundaries. This hike explores the dune and estuary ecosystems of the state park and takes you to wide sandy Imperial Beach.

Directions to trailhead: Border Field State Park is located in the extreme southwestern corner of the United States, with Mexico and the Pacific Ocean as its southern and western boundaries. Exit Interstate 5 Freeway at Palm Avenue, bear west a short distance to 129th Street, south to Monument Road, and west to the park. Follow signs to border monument.

The Hike: A level beach walk 1½ miles north to the Tijuana River and back. For a longer hike, ford the mouth of the river at low tide and continue to the Imperial Beach pier, another 1½ miles. Return the same way. For a full description of this hike see CCT description.

Silver Strand Trail (SD-2)

Coronado to Silver Strand State Park, 6 miles RT

In 1602 Vizcaino named the islands off the coast of Lower California, Los Coronados, after the four Coronado brothers, who were martyred in ancient Rome by Emperor Diocletius. In the 1880s, when the Hotel Del Coronado and its adjoining tent city were built, the beach and town were named after the islands.

North Island is what this extension of the Silver Strand-Coronado peninsula is called today. You may puzzle over this enigmatic name. Its directional name, "north," offers no clue to its meaning. What is it north of? Nor is it an island. Yet Coronado Beach is an unhurried, quiet place, so self-contained that it is commonly thought of as an island. Before the San Diego-Coronado Bay bridge connected the peninsula with the mainland, there was even more of a feeling of isolation.

This hike begins at Coronado City Beach in front of that rambling red-roofed queen of Victorian-era hotels, the Hotel Del Coronado. You'll hike to Silver Strand State Beach along a long sandspit that connects North Island to the mainland at Imperial Beach.

Directions to trailhead: From Interstate 5 in San Diego, take the San Diego-Coronado Bay Bridge exit (State Highway 75). Cross over the bay to the peninsula and make a left on Orange Boulevard, following Highway 75 signs. Park downtown, along Ocean Boulevard or within sight of Hotel Del Coronado. One good way of reaching the beach is through the designated public access through the hotel grounds. If you'd prefer to begin the hike from the south at Silver Strand State Park, take State Highway 75, which serves as a divide between the bay and ocean sides of the peninsula. The park entrance is 1.5 miles north of Imperial Beach.

The Hike: Hike south on the beach in front of the hotel, which is known as Central Beach. North of the hotel is North Beach, south is South Beach. Beyond the hotel, the architecture grows considerably less inspired. You soon reach the Navy Amphibious Base, which occupies much of the northern sandspit. The Navy permits beach hiking, but the base is off limits to the public.

The state park also includes a great deal of San Diego Bay shoreline on the other side of the peninsula. Take one of the pedestrian underpasses beneath busy Highway 75 to the bay side of the park. Here calm, warm water invites a swim.

Return along the Silver Strand the same way.

For further description, see CCT description.

Torrey Pines Beach Trail (SD-3)

Scripps Pier to Flatrock, 11 miles RT

Before or after this day hike, check out the Aquarium Museum at Scripps Institute of Oceanography. In the aquarium all manner of local sea creatures are on display. Underwater video cameras provide views of activity in the nearby marine reserve. Located near the entrance of the Aquarium is a dryland tidepool, where the tide rises and falls in 2-hour intervals. Kelp planted in the pools provide hiding places for bright orange garibaldi, rock fish, and red snapper. Starfish, barnacles, and sea anenomes cling to the rocks. A wave generator simulates surf conditions. The Aquarium is open daily from 9:00 a.m. to 5:00 p.m.

This hike begins at Scripps Pier and passes along Black's Beach, once swimsuits optional, now suits only. After hiking along some spectacular cliffs, you'll arrive at Torrey Pines State Reserve, home to the rare and revered *pinus Torreyana*.

Directions to trailhead: Exit Interstate 5 on La Jolla Village Road, traveling west past UCSD to North Torrey Pines Rd. Turn right, then make a left on La Jolla Shores Drive, following it to the Aquarium-Museum turn-off on your right. Parking is sparse and metered near the Aquarium.

The Hike: As you look south from Scripps Pier, you'll see long and flat La Jolla Shores Beach, a wide expanse of white sand where the water deepens gradually. This is a family beach, one of the safest in the county for swimming. Good diving exists south of the pier where breaks in the continental shelf occur.

Hiking north, the going is rocky at first; the surf really kicks up around Scripps Pier. Soon the beach widens, growing more sandy, and the spectacular, curry-colored cliffs grow higher and higher.

Atop the bluffs is a glider port. Here manned fixed-wing gliders are pulled into the air and ride the currents created by offshore breezes rising up as they meet the cliffs. Brave adventurers strap themselves to hang gliders and leap off the cliffs and, unless the wind shifts, come to a soft landing on the beach below.

The 300-foot cliffs tower over Black's Beach, named for William Black Sr., the oil millionaire who owned and developed most of the land on the cliffs. In the mid-seventies, Black's Beach enjoyed fleeting notoriety as the first and only public beach in the country on which nudity was legal. Called "a noble experiment" by sun worshippers and "a terrible fiasco" by the more inhibited, the clothing optional zone was defeated by San Diegans at the polls.

After passing more handsome bluffs, you'll see a distinct rock outcropping called Flat Rock. For a look at the Torrey pines, ascend the bluff trail to the state reserve.

Return the same way or, if you have transportation waiting, continue another mile to the entrance of the reserve.

Torrey Pines

Del Mar Beach Trail (SD-4)

Train Station to Torrey Pine Reserve, Penasquitos lagoon, 4 miles RT
Train Station to Flat Rock, 6 miles RT

Along Del Mar Beach, the power of the surf is awesome and cliff collapse unpredictable. At this beach, permeable layers of rock tilt toward the sea and lie atop other more impermeable rock layers. Water percolates down through the permeable rocks, settles on the impermeable ones and "greases the skids," an ideal condition for collapsing cliffs. On New Year's Day in 1941 a freight rain suddenly found itself in mid air. Erosion had undermined the tracks. A full passenger train had been delayed and the freight crew of three were the only casualties.

This hike takes you along the beach, visits the superb Flat Rock tidepool area, and detours up the bluffs to Torrey Pines State Reserve. At the Reserve, you'll see those relics from the Ice Age, Torrey Pines, which grow only along the Del Mar bluffs and on Santa Rosa Island, no other place in the world.

Directions to the trailhead: By train: Board the southbound San Diegan in Los Angeles, Santa Barbara, or any other stop along the line and get off in Del Mar. Call Amtrak for fares and schedules. Reservations are usually not necessary.

By car: From Interstate 5, exit on Via de la Valle. Head west to Highway 21, then turn left (south) along the ocean past the race track and fairgrounds to reach the business center. Turn right on 15th street in Del Mar to reach the train station.

> Happily Del Mar is easily reached. Great splendid highways, wide and smooth, pass through it from either San Diego or Los Angeles. Then there are the frequent trains, if you have not the good fortune to possess horses or an automobile. If no other method of transportation is available for your especial need, you might put on your sandal shoes and walk to Del Mar as good old Father Junipero Serra did many a time before you...
> —*Sunset*, 1912

The Hike: From the station, cross the tracks to the beach and begin hiking south. With the high cliffs on your left and pounding breakers on your right, you'll feel you're entering another world. Follow the sometimes wide, sometimes narrow beach over sparkling sand and soft green limestone rock. The greenish beds are part of the Del Mar Formation. Above is cavernous, weathered Torrey Sandstone. As you hike up coast, keep an eye on the cliff base. You may see fossil fragments in places where resistant sandstone forms an outcropping ledge.

After two miles of beachcombing you'll reach the north end of the Reserve and Penasquitos Lagoon, a salt water marsh patrolled by native and migratory waterfowl. Another mile south you'll see a distinct flat outcropping named appropriately enough, Flat Rock. Common tidepool residents, housed in the rocks and base of the bluff include barnacles, mussels, crabs, and sea anenomes.

Just north of Flat Rock, a trail ascends the bluffs of the State Reserve. The first hundred yards of the unsigned trail is tentative and precarious, but it's easy walking once it begins climbing the sandstone bluffs. The Reserve is home of the Torrey Pine, which occupies the bold headlands atop the cliffs. Clinging to the crumbling yellow sandstone, these rare and graceful trees seem to thrive on the foggy atmosphere and precarious footing.

The Reserve is filled with yellow coreopsis, lupine, Indian paint brush and some interpretive nature trails, complete with self-guiding leaflets. The Guy Fleming loop trail, a short 6/10 of a mile, points out the Torrey pines and takes you to South Overlook, where you might glimpse a migrating gray whale.

Return the same way.

Three Lagoons Trail (SD-5)

Leucadia State Beach to South Carlsbad State Beach, 6 miles RT
Leucadia State Beach to Carlsbad State Beach, 10 miles RT

Leucadia, like most of north San Diego County, is surfer country. However in Leucadia, the noisome public beach atmosphere is conspicuously absent and the communion between surfer and wave is personal, even spiritual. At the base of the cliffs stretch out Grandview and Beacons Beaches, preferred by surfing connosieurs. A silence begins, a quiet appreciation of sun and surf.

This hike heads north from Leucadia Beach, visits three lagoons, Batiquitos, Agua Hedionda, and Buena Vista, and ends up at Carlsbad State Beach. The beaches, beach towns and lagoons along this hike are quiet and peaceful; they were bypassed when Interstate 5 opened in 1966 and routed traffic inland.

Directions to trailhead: From Interstate 5, exit on Leucadia Boulevard and head west to Pacific Coast Highway. Go north on PCH to Grandview and make a left to the parking lot. You'll find a stairway down to the beach. Mass Transit: North County Transit NCTD bus #301 and #361.

The Hike: Hike north on Leucadia State Beach, a wide sandy beach backed by bluffs. Good swimming and surfing abound here. In a half mile you'll reach Batiquitos Lagoon, an unprotected area serving as a refuge for endangered birds.

A mile past the lagoon you begin crossing South Carlsbad State Beach. A large developed campsite is atop the sandstone bluffs. The trail continues over clean sand and rock beach to Carlsbad State Beach and the town of Carlsbad.

Pioneer John Frazier homesteaded here in 1883 and soon it was known as Frazier's Station. Digging a well, Frazier discovered mineral water. Analysis determined this water to be equal in mineral content to that of Karlsbad, Czechoslavakia.

Are you ailing, are you failing,	It is Carlsbad, bonny Carlsbad,
Have you ilk you cannot tell?	And upon its sparkling brink,
There is healing past revealing	Hygeia sits forever smiling—
In the waters of the well.	And she bids you come and drink.

—Anonymous Carlsbad resident, 1890

On the state beach, people row, surf, dive, swim, and build sand castles. You'll find good surfing and swimming between the two jetties. Watch your step—sting rays, sea urchins, and rip currents lurk here.

Across the highway from the state beach is Agua Hedionda (Spanish for "Stinking Water") Lagoon, named for the sulphur spring at Carlsbad. Early Californios often applied Agua Hedionda to California springs and ponds. When the Yankees came, they usually changed the name to "Sulphur Springs" and enriched themselves by developing them as health resorts.

At the lagoon is a San Diego Gas and Electric power plant. The company avoided the stinking water designation and chose "Encina" for the plant's name. Encina translates to oak, of which there are none native in the vicinity. The calm, no-longer-odiferous waters of the lagoon make a popular spot for boating and fishing. Fishermen cast for corbina and croaker along the Encina Fishing Area at the mouth of the lagoon.

You may return the same way or continue another two miles to the other end of Carlsbad and yet another lagoon, Buena Vista. This lagoon was initially formed as an estuary when the mouth of Buena Vista Creek was drowned by a rising sea level. It's now separated from the ocean by a broad sand bar. Fishermen try their luck for bass, bluegill and catfish.

Agua Hedionda Lagoon

Bayside Trail (SD-6)

Old Point Loma Lighthouse to Cabrillo National Monument Boundary, 2 miles RT

Cabrillo National Monument on the tip of Point Loma marks the point where Portuguese navigator Juan Rodriguez Cabrillo was the first European to set foot on California soil. He landed near Ballast Point in 1542 and claimed San Diego Bay for Spain. Cabrillo liked this "closed and very good port" and said so in his report to the king.

Point Loma Lighthouse, built by the federal government, first shined its beacon in 1855. Because fog often obscured the light, the station was abandoned in 1891 and a new one built on lower ground at the tip of Point Loma. The 1891 lighthouse is still in operation today, but manned by the Coast Guard. The 1855 lighthouse has been wonderfully restored to the way it looked when Captain Israel and his family lived there in the 1880s.

The Bayside Trail begins at the old lighthouse and winds past yucca and prickly pear, sage and buckwheat. The Monument protects one of the last patches of native flora, a hint at how San Diego Bay must have looked when Cabrillo's two small ships anchored here.

Directions to trailhead: Cabrillo National Monument is open daily from 9:00 a.m. to 5:15 p.m. (8:30 a.m. to 7:45 p.m. June 25 to September 4). Take Harbor Drive West to State Highway 209. Turn left and follow the signs to the park. Mass Transit: SDTC #6.

The Hike: Before embarking on this hike you may want to obtain a trail guide at the Visitors Center. The guide describes the coastal sage and chaparral comunities and local history.

The first part of the Bayside Trail winding down from the lighthouse, is a paved road. At a barrier, you'll bear left on a gravel road, once a military patrol road. During World War II the Navy secreted bunkers and searchlights along the coastal bluffs. Now, in spring, the slopes are covered with black-eyed Susans, Indian paint brush, sea dahlias, monkey flowers.

The Bayside Trail provides fine views of the San Diego Harbor shipping lanes. When Navy ships pass, park rangers broadcast descriptions of the vessels. Also along the trail is one of California's most popular panormaic views: miles of seashore, 6,000-foot mountains to the east and Mexico to the south.

The trail deadends at the park boundary.

Return the same way.

Sunset Cliffs Trail (SD-7)

**Eucalyptus Grove to Fort Rosecrans,
2 miles RT**

Sunset Cliffs, on the ocean side of Point Loma, are projections of sculpted sandstone, which frame small pocket beaches. Divers and (experienced) surfers come here as well as sunset fans who marvel at the brilliant light reflecting off the golden cliffs. The cliffs begin just south of Ocean Pier and extend to Fort Rosecrans Military Reservation.

This trail, not for acrophobics, leads along the edge of the Sunset Cliffs, providing a bird's eye view of the surfing action below and a good view of the coastline. The trails along the cliff are highly eroded and can be dangerous.

The Hike: The Sunset Cliffs Trail heads south from the eucalyptus grove adjoining the parking area. Although an Australian import, the eucalyptus seems to be a natural element here. No other imported tree appears more at home along the California coastline. Their blue and white foliage, peeling bark, and wild fragrance get your hike off to a good start and suggest a picnic spot for return.

Directions to trailhead: Take Sunset Cliffs Boulevard to its end, bearing right onto a dirt road into Sunset Cliffs Park. Mass Transit: SDTC bus #35.

The trail starts along the cliffs, but soon detours inland around a baseball diamond. Continuing past a school, the trail becomes catch-as-catch-can. One trail follows the cliff edge, rising up and down and over the eroded sandstone. Another leads through high mustard, interspersed with seasonal wildflowers.

Turn around before trespassing on Fort Military Reservation property. If it's low tide, you can return via the beach.

North County Trail (SD-8)

Del Mar to Oceanside, 25 miles one way

Twenty-five miles of obstruction-free beach await the hiker in north San Diego County. Stretching between Del Mar and Oceanside is a coastline of black and white sand beaches, backed by bluffs and dotted with lagoons. The towns and beaches here are quiet and give the illusion of being off the beaten track. Until 1964, it was busier because Highway 101, the main thoroughfare between Los Angeles and San Diego, sent motorists whizzing along the coast. However, when Interstate 5 routed long distance travelers inland, the Coast Highway was left to residents and beach goers.

To hikers, the towns of Del Mar, Solana Beach, Cardiff-by-the-Sea, Encinitas, and Leucadia all look alike, though residents would be quick to dispute this and point out the superiority of their town and beach. There are a half dozen state beaches along this hike and two of them, San Elijo and South Carlsbad, have camping. Amtrak stops at Del Mar and Oceanside so any way you hike it, you'll have a nice two- or three-day weekend in North County.

Directions to trailhead: Take Amtrak's southbound San Diegan from L.A. or any other point along the line and get off at Del Mar Station. The train station in Oceanside is located at 117 N. Cleveland.

The Hike: Consult the CCT description of North County for an account of the hike from Del Mar to Oceanside. Remember, the only two campgrounds enroute are at San Elijo and South Carlsbad State Beaches, so pace yourself accordingly.

Orange County

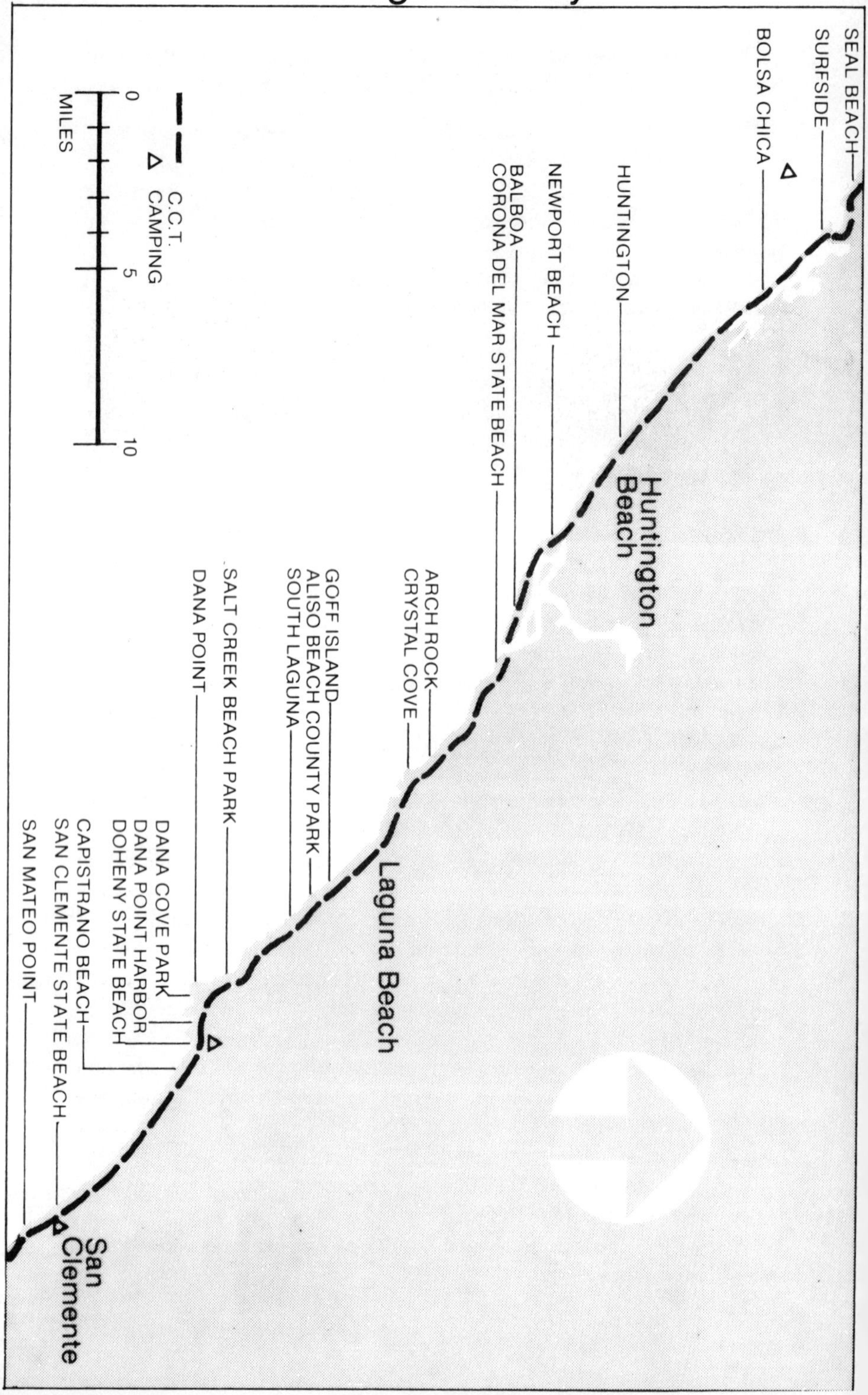

MILES 0 — 5 — 10

- - - C.C.T.
△ CAMPING

SEAL BEACH
SURFSIDE
BOLSA CHICA △
HUNTINGTON
NEWPORT BEACH
BALBOA
CORONA DEL MAR STATE BEACH
ARCH ROCK
CRYSTAL COVE
GOFF ISLAND
ALISO BEACH COUNTY PARK
SOUTH LAGUNA
SALT CREEK BEACH PARK
DANA POINT
DANA COVE PARK
DANA POINT HARBOR
DOHENY STATE BEACH △
CAPISTRANO BEACH
SAN CLEMENTE STATE BEACH
SAN MATEO POINT

Huntington Beach
Laguna Beach
San Clemente △

Orange County

Since before the turn of the century when inland farmers took their families to Arch Beach to camp out and cool off, the Orange County coast has served as a resort area for Southland residents. Balboa Island and Tin Can Beach, San Juan Capistrano and Laguna have been popular destinations for generations.

Although it shares a single coastal plain with neighboring Los Angeles County, Orange County retains a distinct shoreline identity as a boating community (Newport Harbor) a surfing community (Huntington Beach) and an art colony (Laguna Beach). The pleasant climate and varied Pacific shoreline make the coast here much sought-after for recreation and residence.

From San Onofre to Seal Beach, CCT travels over a dozen major public beaches and numerous pocket beaches. The sandy and cobble beaches are clean, quiet and cheerful, considering the heavy use they receive.

The Orange County shoreline extends 45 miles in a northwesterly direction from San Mateo Point to the San Gabriel River. Sandy Doheny and San Clemente State Beaches at the south end of the county offer good camping. Between Dana Point and Laguna the shore is rockier, interspersed with pocket beaches backed with 60 to 200-foot cliffs of marine sedimentary siltstone, sandstone and shale. From the Newport Harbor entrance to Long Beach, the shoreline is wide sandy beach broken by marshes and rocky points. Piers and jetties deflect the waves for surfers. The Newport-Balboa area is the most commercial part of the county's coastline and is dominated by the huge Newport marina complex.

The Santa Ana Mountains may be the most overlooked and under-utilized recreation area in Southern California. Stretching the entire length of Orange County's eastern perimeter, the Santa Anas roughly parallel the coast. This coastal range is only twenty miles inland and the western slopes are often blanketed with fog. These mountains are round, brushy, inviting. The coast has a cooling influence on them. Except for the dead of summer, most days offer pleasant hiking.

The orchards that gave Orange County its pastoral name are mostly gone now, but the coast ranges occupying half the county still afford invigorating vistas as do the bluffs overlooking the ocean. Although many homes crowd the surf, much of the coastline is still alluring.

CCT at a glance

TERRAIN: Mostly sand beach. Considering the county's huge population and rapid growth, the beach is more natural than you might expect.

OBSTRUCTIONS: Homes built close to the water create *de facto* private beaches. Problem areas include Dana Point, Laguna, Irvine Cove.

TRANSPORTATION: Orange County Transit District: (714) 636-RIDE. Amtrak stops in Santa Ana and San Clemente.

CAMPGROUNDS-ACCOMMODATIONS
Huntington City Beach (September 15-May 31 only)
Newport Dunes Aquatic Park (discourages walk-in campers)
Doheny State Beach
San Clemente State Beach
No hostels, but many motels.

CCT: ORANGE COUNTY

CCT rounds San Mateo Point into Orange County amidst dire warnings about private property and follows the sandy shoreline of exclusive Cotton's Beach. If you hike too close to the expensive homes and manicured lawns, you might be shouted at, but passage is no problem on the lower beach. The railroad tracks run right along the beach from here to Dana Point. (When a train comes by it's actually more of an event than a nuisance.)

Three quarters of a mile from the point is San Clemente State Beach. A number of bluff trails take hikers to the camping area on the cliffs, 100 feet above the beach. The campground is a few hundred yards from former President Nixon's Western White House. The campground includes a special hiker-biker camp.

CCT journeys up-coast on the sandy state beach, where you'll find great fishing and swimming. In 1½ miles you'll reach San Clemente Municipal Pier, a popular fishing spot. North of the pier extends San Clemente Beach, scene of fierce volleyball competition.

CCT continues to Capistrano Beach, which bounds the old mission town of San Juan Capistrano. In 1887 during the big land boom in Southern California, a high-class sub-division called San Juan By-the-Sea was laid out on the mesa land near San Juan Creek. It went bust, but was revived in 1925 as Capistrano Beach. Be sure to walk to town for a look at the old mission and to inquire if the swallows have returned.

CCT continues on the sandy strand of Doheny Beach, formed through the efforts of San Juan Creek, which delivers to Pacific waves the material needed to build the strand. The state beach park, which opened in 1931, has 115 campsites. It's not the most quiet park, due to its close proximity to the highway, but the sites on the beach side of the campground will be fine for tired hikers. There's a special hiker-biker campsite.

A short distance past Doheny State Beach, hikers cross a shallow lagoon at the mouth of San Juan Creek. It's easy to cross, particularly at low tide. Another quarter mile brings you to a jetty marking the entrance of Dana Point Harbor. Bear north to Del Obispo Street, then go left. Improvise a route through the modern, antiseptic marina.

> Down this height we pitched the hides, throwing them as far out into the air as we could; and as they were all large, stiff, and doubled, like the cover of a book, the wind took them, and they swayed and eddied about, plunging and rising in the air, like a kite when it has broken its string.
>
> —Richard Henry Dana, *Two Years Before the Mast*

In 1835, Richard Henry Dana, a young sailor on the *Alert* tossed cowhides off the point which now bears his name. His shipmates on the beach below, balanced the hides on their heads and walked them to the waiting ship. A few of Dana's tosses went astray and his shipmates lowered him by rope to retrieve a half dozen hides which had lodged against the precipitous bluffs. Later, Dana wrote the classic book about life on the high seas called *Two Years Before the Mast*.

Richard Henry Dana Monument

If you'd like, walk the paths on the landward and seaward sides of Dana Island. Fishermen work the spray-soaked jetties and the cove at the west end of the harbor. A statue honors Richard Henry Dana. Extending toward Dana Point along Del Obispo Road is Golden Lantern Lookout Park, a fine picnic ground. CCT arrives at the foot of some handsome bluffs, seaward and west of the harbor. The cliffs here and those opposite the picnic area consist of coarse breccia and the sandstones and shales of the Capistrano Formation.

Tiny Dana Cove is a prime area for tidepool exploring, diving, and snorkeling. To reach the area, you must scramble over the breakwater rocks, past a small beach. This area is preserved in one of Orange County's marine life refuges. You may continue about a quarter mile up-coast from the park, until the waves dash against the cliffs and the route becomes impassable.

In order to round Dana Point, you must ascend the bluffs behind Dana Cove Park. Follow Cove Road, which is more of a trail, because it has been closed to autos due to landslides. As you ascend the steep road, you'll get a terrific view of Dana Harbor.

Once atop the bluff, continue on the dirt road, keeping the ocean on your left and ignoring the wall of houses blocking your access to Dana Point. Scenic Drive on your left soon becomes Dana Strand Road. CCT reaches this road as residential housing grows sporadic. You'll see a number of eroded and unstable surfer trails switchbacking down the bluffs. Carefully choose a route and descend cautiously to the west side of Dana Point. When you reach the shoreline once more, pause to appreciate the severely deformed light-colored shale and Monterey sandstone beds of the point.

Sandy Nigel Beach spreads before you. The beach has a secluded feeling, even though the Laguna Nigel housing development is situated on the low bluffs above the beach. A small headland divides the beach in two. There's good surfing on the offshore reef; the waters are shallow and provide large obliquely breaking swells.

Beachwalk to Salt Creek Beach Park, named for the salt grass that grows farther inland. The mouth of the creek is a cement channel. Beyond is a YMCA boating camp. Hikers may continue another quarter mile up-coast but will run into a mess of barbed wire and cement, a residential Maginot Line. This will be the first of several times in Orange County that hikers are denied passage. Hikers have three options here:

1) Retreat back to Salt Creek Beach Park, follow the park road a quarter mile up the bluffs along the park road to Pacific Coast Highway, resume walking north.
2) Hike up the Salt Creek spillway, the site of much construction, to Pacific Coast Highway.
3) Take surface streets through the walled development of Monarch Bay and suffer the hostilities of guards when you emerge at one of their outpost gates along the coast highway, resume walking north. All routes continue along Highway 1, reaching South Laguna via a posted bike route.

One minor consolation of being forced inland is that 50 yards south of the Monarch Bay Guard House (where you'll emerge if you choose Option 3) is California Historical Marker 189 commemorating Dana Point. It's ironic discovering Dana so far from his beloved sea. Perhaps some day entrenched private property interests will be overcome and CCT hikers and Richard Henry Dana allowed near the water. At present, hikers will be the only persons paying homage to Dana, since the historical marker is overgrown by brush and overlooked by motorists whizzing by on Highway 1.

After passing the gated, guarded community of Three Arch Bay, you'll enter the commercial district of South Laguna Beach. CCT continues to a series of pocket beaches. Depending on the tide and your mountaineering skill, you may clamber from one to another, but be warned that private property and steep rock (the former more of a threat than the latter) make this risky business. The first really convenient beach access is at Table Rock Park, on Table Rock Rd., West of PCH.

Beyond Table Rock (fine picnicking as its name suggests) is Coast Royal County Beach, a mixed sandy and rocky shore. In 1906 a shyster named Horace J. Pullen acquired partial interest in the shoreline here and without the principal owner's permission, laid out a subdivision he named Coast Royal. His unwitting and irate partner, Miss Blanche Delphi, in Europe at the time, heard of his activity and filed suit. Los Angeles newspapers branded Coast Royal a swindle and the project died. In the Twenties the subdivision was revived and developed into the exclusive community it is today.

As CCT rounds rocky Aliso Point, you'll hike over the storm benches exposed at low tide. The rock and sea cliff above is San Onofre breccia, one of the most spectacular sedimentary deposits in Southern California. The breccia consists of angular fragments up to several feet in diameter, of a wide variety of metamorphic rock, mostly greenish to black in color with a wide range of textures. You might find fragments of a fine-grained rock of bluish-lavender hue known as glaucophane schist. The nearest exposures of bedrock like the fragments in this breccia are on Catalina Island and Palos Verdes Hills. Geologically curious hikers will pause at Aliso Point for profound speculation on plate tectonics and continental drift.

After rounding the point, you'll arrive at Aliso Beach County Park, one of the nicest Laguna cove beaches. It's a popular surfing spot. A long fishing pier juts out from the south end of the beach.

Three quarters of a mile up-coast from the park is a rocky peninsula known as Goff's Island. Early in the 1870s, the four Goff brothers settled here. Forty yards offshore was a half acre island rising to a height of 45 feet above sea level. It was thought to be part of the Goff property and acquired its name, but by Congressional proclamation in 1931 it was declared to be public domain and set aside for its scenic beauty. The rocky projection, now connected to the mainland, has been used in several motion pictures. Sometimes it's referred to as Treasure Island.

Around the rocky peninsula is the mouth of tiny Hobo Canyon. In the old days the canyon is said to have been a favored rendezvous for those unable to afford the high-priced accommodations in the Laguna art colony.

San Onofre breccia and private property deny passage a half mile up-coast at Sugar Loaf Point. You'll have to detour inland a few blocks through a residential area to Pearl Street at the south end of Laguna Beach proper. Once back on the shoreline, you soon encounter a snag of metamorphic rock projecting to the surf line called Halfway Rock; referred to thusly, because the brown barrier is halfway between Laguna Beach and Arch Beach.

It's easy walking to main Laguna Beach. The long beach with its boardwalk borders the center of town. At the north end of the beach Laguna Canyon Road meets Pacific Coast Highway. A short distance up the road the world famous Laguna Art Festival and Pageant of the Masters takes place. Laguna Canyon is the longest canyon in the San Joaquin Hills and was called Canon de las Lagunas (Canyon of the Lagoons), during

Spanish days. In 1887 George Rogers purchased the land at the mouth of the canyon for $1,000 and laid out a subdivision he called Laguna Beach. The town was incorporated in 1927 and early art colony residents dubbed themselves the Lagunatics.

North of Laguna Canyon, CCT grows increasingly rocky; passage is difficult at high tide. On the bluffs above the shoreline is grassy Heisler Park, where oldsters enjoy lawn bowling. The Laguna Art Museum is perched here and it is a good picnic site. CCT passes several pocket beaches, first Rock Pile Beach, then Picnic Beach. If the tide is high, you can retreat up any of a half dozen stairwells to Cliff Drive, and a shoreline park, and simply follow the park up-coast. If the tide is low, continue up-beach past Diver's Cove. The cove is part of Glenn E. Vedder Ecological Reserve offshore, a popular snorkeling area.

CCT continues on rocky and sandy beach to Twin Points, a rocky headland forming the southern end of Crescent Bay. A sandy beach extends a quarter mile to Two Rocks Point. Near the end of Crescent Bay Beach you'll have to bear inland at Cliff Drive and return one block to Pacific Coast Highway. CCT follows the highway up-coast for a dull mile and half because there's no beach access through the walled, guarded communities of Emerald Bay and Irvine Cove. Cliffs backing the irregular shore of Irvine Cove and the peninsula-like point enclosing Emerald Bay on the northwest are composed of resistant andesite. Atop the cliffs are resistant property holders.

Beyond Irvine Cove gatehouse, and just outside the Laguna Beach City limits, is Abalone Point, a rocky promontory made of eroded lavas and other volcanic material distributed in the San Joaquin Hills. It's capped by a grass-covered dome rising two hundred feet above the water.

A few hundred feet up-coast from the point you'll encounter a trailer park at the mouth of Moro Canyon. A pedestrian access-way allows you through the park back to te beach. The name is a misspelling of the Spanish word morro, meaning round, and describes the round dome of Abalone Point.

Hike up-coast on sandy El Moro Beach, which is sometimes beautifully cusped. The beach is remarkably undeveloped and pristine. The beach and cliffs north to Arch Rock near Corona Del Mar were recently purchased by the state from the Irvine Corporation. Trails lead up the bluffs where horseback riding is popular. (Hikers may walk atop the bluffs a few miles if preferred.) CCT passes Reef Point, a low point and cluster of rocks near the shoreline and soon reaches Crystal Cove, site of a few simple beach cottages. Cove is something of a misnomer, for this beach shows no indentation of any kind.

A little south of Corona Del Mar city limits, the beach begins to narrow and grow rocky. Hikers squeeze closer to the white, thin bedded shale cliffs. Through well-named Arch Rock, you can sight sailboats tacking out of Newport Harbor. Beyond is a snug pocket of sand, Little Corona Del Mar Beach. Excellent tidepools abound here; offshore is the Newport Marine Life Refuse, a popular diving spot.

Some rock scrambling brings you to "Big" Corona Del Mar Beach, large and sandy, with all the facilities common to a popular sunbathing beach. On the west end of the beach is Newport Harbor's East Jetty, scene of surfer and diver activity.

Newport Bay was first used as a seaport in the 1860s. Lower Newport Bay, which you'll be hiking around ,was dredged in the 1930s to its present configuration with seven islands and a 20-foot main channel. Home of 8,000 pleasure craft, the bay ranks among the world's largest yacht harbors.

CCT follows the jetty briefly, then ascends the stairway up to Ocean Drive. Because of private property and the hands-off nature of the bay, you can't walk along the marinas; CCT follows first Ocean, then Bayside Drive. There's access to the water, particularly at China Cove Beach, but the labyrinthine layout of the marina complicates things. You'll spot a pier and

Balboa Ferry

the tile-roofed Kerkhoff Laboratory belonging to California Institute of Technology, which conducts oceanic research here.

CCT follows Bayside Bay around Newport Harbor to a jumbo intersection with Jamboree Road. Here you'll turn left and walk over the bridge to Balboa Island, taking Marine Avenue through the center of town to the bay side of the island.

Those thinking about camping will continue following Bayside Drive at the jumbo intersection, past Pacific Coast Highway to Newport Dunes Aquatic Park. It's a 15-acre lagoon with swimming and camping facilities. It's mostly oriented to vehicle camping, but there's tenting on the beach.

Well worth a look (and hike; see Newport Back Bay trail) is Upper Newport Bay, a state operated Ecological Reserve, bounded by eel grass and pickleweed, refuge for more than 150 species of birds, including Pacific Flyway migrants.

CCT follows the scenic walkway on the south side of the island. Balboa Island had its beginning in 1906 when W.S. Collins dredged bay mud onto a sand flat that appeared in Newport Bay during low tide. He subdivided the island and by 1914, more than one-half the 1300 lots were sold.

CCT reaches the Balboa Island Ferry Landing at Agate Avenue. Here you catch the ferry, which runs frequently and is a bargain at 15¢ a hiker. Ferry services was innaugurated in 1907 by a genial black man named John Watts, who encouraged his open launch, the *Teal,* with great draughts from an oil can. Today's small auto ferrys make the three hundred yard voyage in under five minutes.

The ferry leaves you at the foot of Palm Street near the historic Balboa boardwalk. Remodeling and new construction have obliterated most of the early building fostered by electric railway service from Los Angeles begun in 1905. Still standing is a landmark pavilion and amusement zone.

CCT follows Palm Street across Balboa Boulevard to Balboa Beach. A short distance down-coast is Balboa Pier, site of the first water-to-water flight in 1912. Glenn L. Martin flew a hydroplane from the waters here to Avalon Bay at Catalina Island.

CCT proceeds up-coast along 5¼ mile long Newport Beach, held together by a series of groins. In 1½ miles you reach Newport Pier, a historical landmark. A wharf built here in the 1890s served as a shipping point for Orange County produce and gave rise to the city of Newport Beach.

Beyond the pier the beach narrows a little and CCT crosses Santa Ana River County Beach. At the river, CCT follows the east bank inland a few hundred yards up and down sand dunes to the Highway 1 bridge, and crosses the bridge to Huntington State Beach. Formerly, the Santa Ana River emptied into the ocean at Newport Bay, but was diverted to shore at this point in order to reduce silting after the bay became a great yacht harbor. The Santa Ana River Trail is primarily a scenic bikeway which follows first the west side of the river, then the east, and ultimately ends up in Yorba Linda.

Huntington Beach was once called Shell Beach because millions of small bean clams, *Donax*, were washed up on its sands. In recent years Pismo clamming has undergone something of a revival here. Pacific Electric Railroad magnate Henry E. Huntington was active in the area's development at the turn of the century and gave the new town his name. Three-mile-long Huntington State Beach hosts more than a million visitors annually.

Huntington City Beach is a continuation of the same sandy beach. It's best known as the site of international surfing competitions. An 1800 foot pier, built in 1914, bisects the beach. The pier has a bait and tackle shop and is floodlit at night for fishing and surfing.

Huntington Pier

Note to campers: Between September 15 and May 31, camping is allowed at Huntington City Beach for a fee. The campground is mostly oriented toward RV's, but will sometimes grudgingly accommodate hikers. Entrance is off Pacific Coast Highway, opposite Lake Avenue. Information and reservations: City of Huntington Beach—Sunset Vista, P.O. Box 190, Huntington Beach, CA 92648 (714) 536-5281, -5280.

Beyond Huntington Pier stretches 6-mile long Bolsa Chica State Beach. The southern end is undeveloped and is a continuation of the long wide sandy beach passing through the community of Huntington Beach. The sandy bottom is great for swimming. You'll find good surfing and surf fishing. Above the beach is Huntington Beach Mesa, or "The Cliffs" as the surfers call it.

Until Bolsa Chica Beach came under the administration of the state in 1961, the long sandy shoreline here was popular with picnickers and campers. No effort was made to keep the beach clean, so the popular name of Tin Can Beach was applied.

Opposite the state beach is Bolsa Chica Ecological Reserve. This 1800-acre marshland is somewhat degraded, because for years its ocean outlet was dammed up. The state has restored part of the marsh bordering the highway, but the rest belongs to an oil company and its future is very much in doubt.

In the 1920s oil was discovered at Bolsa Chica. The oil field is currently one of the largest in Southern California, with over 200 active wells. They stand shoulder to shoulder on the cliffs above the beach.

North of Bolsa Chica, CCT crosses Sunset Beach, which occupies the sandspit between the ocean and tideland. The community is one of several that developed along the ocean front when Henry Huntington extended his Pacific Electric railroad down the coast from Long Beach to Balboa in the early 1900s.

Sunset Beach, one of Orange County's most popular beaches, is backed by low sand dunes and ice plant. Opposite Sunset Beach, on the other side of Highway 1, is Hungtington Harbor, a private harbor adjacent to expensive waterfront homes. It has little interest to the hiker.

As the jetty protecting Anaheim Bay comes into view, CCT arrives at Surfside Beach. Surfside is a small coastal resort adjoining the Los Alamitos Naval Weapons Center. Because there's no trespassing onto Naval property, you must leave the beach here through the Anderson Street gate of the private Surfside community and walk north along Pacific Coast Highway.

After one mile, you'll pass by Seal Beach National Wildlife Refuge, 1200 acres of salt marshes which belong to the Navy. Anaheim Bay on the west side of the highway was developed as a seaport for Orange County, prior to the emergence of San Pedro. The swampy sloughs here served as a shipping point for Anaheim farmers. Boats carrying produce would sail out on high tide to anchored ships and return with needed building materials.

During World War II, the Navy converted the outer bay into an ammunition depot, dredging an entrance channel and building stone jetties. Today the Naval Weapons Station handles conventional and "special" (nuclear) arms.

At Seal Beach Boulevard, CCT goes west three blocks, returning hikers to the beach. Once called Bay City, the town here was christened Seal Beach when incorporated in 1915 because of the large numbers of harbor seals at the mouth of the San Gabriel River. Like many other coastal towns, Seal Beach has expanded with industry and suburbs and retains almost none of its original resort orientation.

CCT passes Seal Beach pier, which divides the beach in two and is a focal point for surfers and fishermen. It's floodlit at night.

CCT arrives at the low dunes at the mouth of the San Gabriel River. Upriver, power plants located on the east and west sides, discharge hot water into the river channel. The discharge, often 20 degrees warmer than normal sea water, is often seen "steaming." The warm water is welcomed by surfers riding waves along the jetty, particularly during the winter months; it also attracts both desirable and undesirable species of sea life.

CCT heads a quarter mile inland along the east bank of the San Gabriel River, where fishermen try their luck. Our route crosses the river at Marina Drive Bridge and enters Los Angeles County.

Aliso Beach Trail (0-1)

Aliso County Park to Goff Island, 2 miles RT

Goff Island, a short mile upcoast from Aliso County Park, was named after four brothers, who settled along the coast in South Laguna in the 1870s. The rocky half-acre island rising 45 feet above the breakers, was later connected to the mainland. In 1931, Congress declared it to be public domain and set it aside.

This hike travels Aliso Beach, the wide sandy beach at the mouth of Aliso Canyon, passes through handsome Arch Rock, and visits historic Goff Island.

Directions to trailhead: Aliso Beach County Park is right off Highway 1, three miles south of the city of Laguna Beach.

The Hike: Begin at the sport fishing pier and walk up the sandy beach. Occasionally there's water in Aliso Creek, but it is always easy to ford. The Orange County Directory of 1895 describes Aliso Beach as "campsite and seaside resort for farmers, a short distance south of Arch Beach."

Aliso Beach soon narrows and becomes rocky. You'll hike below some expensive homes perched on the cliffs and pass through a natural rock arch. In 1887 Hubbard and Henry Goff laid out a subdivision a little to the south which they named Arch Beach after this rock. The subdivision failed; the name and rock remain.

The peninsula of Goff Island is a beautiful meeting of rock and surf. You'll see the remains of an old boat landing. Just up-coast is Hobo Canyon, where struggling artists and assorted itinerants who couldn't afford Laguna Beach accommodations (even in the old days the art colony was expensive), camped out.

You may return the same way or press on over rocky beach for another 1/3 mile to Sugarloaf Point, where the tides and private property make it impossible to continue.

Crown of the Sea Trail (0-2)

Corona Del Mar Beach to Arch Rock, 2 miles RT
Corona Del Mar Beach to Crystal Cove, 4 miles RT
Corona Del Mar Beach to Abalone Point, 7 miles RT

In 1904 George Hart purchased 700 acres of land on the cliffs east of the entrance to Newport Bay and laid out a subdivision he called Corona Del Mar ("Crown of the Sea"). The only way to reach the townsite was by way of a long muddy road that circled around the head of Upper Newport Bay. Later a ferry carried tourists and residents from Balboa to Corona Del Mar. Little civic improvement occurred until Highway 101 bridged the bay and the community was annexed to Newport Beach.

This hike explores the beaches and marine refuges of "Big" and Little Corona Del Mar Beaches and continues to undeveloped Irvine Beach, recently acquired by the state.

Directions to trailhead: From Pacific Coast Highway in Corona Del Mar, turn oceanward on Marguerite Avenue, traveling a few blocks to Corona Del Mar State Beach. Mass Transit: OCTD #1 and #57.

The Hike: Begin at the east jetty of Newport Beach, where you'll see sail boats tacking in and out of the harbor. Snorkelers and surfers frequent the rocky area of the jetty. Proceed down-coast along wide sandy Corona Del Mar State Beach. This well-developed beach includes picnic tables, outdoor showers and raft rentals and is popular with tanners and swimmers.

The beach narrows as you approach the cove that encloses Little Corona Del Mar Beach. Snorkeling is good beneath the cliffs of "Big" and Little Corona Beaches. Both areas are protected from boat traffic by kelp beds and marine refuge status.

A mile from the jetty you'll pass well-named Arch Rock, which is just off shore and can be reached at low tide. The beach from Arch Rock to Irvine Cove, 2½ miles to the south, was recently purchased by the state from the Irvine Corporation. Trails lead up the bluffs to a stable that rents horses. A mile's hike down the undeveloped beach brings you to Crystal Cove, a cluster of cottages between Pacific Coast Highway and the ocean at the mouth of Los Trancos Canyon. Don't look for a cove; the name is a misnomer, for there's no coastal identation here at all.

If you wish, continue another mile and a half to sandy El Moro Beach, which is sometimes beautifully cusped. Ahead is Abalone Point, a rocky promontory capped by a grass-covered dome rising 200 feet above the water. On your way back you may walk some of the distance atop Irvine Beach bluffs, or return the same way along the beach.

Newport Trail (0-3)

Newport Pier to Balboa Pier, 3 miles RT
Newport Pier to Jetty View Park, 5 miles RT

At the turn of the century the McFadden brothers built a wharf at Newport, ushering in an era of commerce and providing a shipping point for Orange County farmers. A historical landmark commemorates the original wharf.

This hike travels east along the sandspit which shelters Newport Bay from the ocean. You'll pass Balboa Beach, where a side trip may be taken to Balboa Island, and continue to Jetty View Park at the entrance of Newport Harbor.

Directions to trailhead: Follow Newport Boulevard (highway 55) to its end at Balboa Boulevard. Mass Transit: OCTD #1, #53, #65.

The Hike: Hike down-coast on Newport Beach. Before Newport Pier was constructed, the long strand along the ocean was known as Sand Beach. The beach and sea floor slope steeply, generating large waves and severe erosion rates. Newport Beach is held together by a series of small groins. Sand washed from the beach is believed to collect in the Newport Submarine Canyon, which trends south from the pier.

Hikers may detour inland a hundred yards from Balboa Pier to a landing at Palm Avenue and for 20¢ be ferried over to Balboa Island. Ferry service began in 1907. With the clang of bells, a one-cylinder engine powered the open launch the *Teal* on its three-hundred-yard voyage from Newport to Balboa. You can make the same voyage that delighted tourists eight decades ago and spend some time exploring the bay front boardwalk and small beaches of the island.

A short distance down Balboa Beach, you'll reach Balboa Pier. The mile of beach from pier to the west jetty is generally sparsely populated. Jetty View Park is a shaded park between Balboa Beach and the jetty. Before the jetty was improved and extended, entering Newport Harbor during storms could be dangerous. One Sunday morning in June 1925, a forty-foot fishing boat, the *Thelma*, with 17 people on board, arrived in a storm as waves crashed over the jetty. The boat capsized. As luck would have it, Hawaiian swim champion and one of the founding fathers of surfing Duke Kahanamoku was surfing the big waves at Corona Del Mar Beach. Duke and his surfing buddies paddled out and managed to bring ashore all but five of the victims.

After yacht watching, return the same way.

Back Bay Trail (0-4)

3½ miles RT

In 1974, Orange County and the Irvine Company reached an agreement calling for public ownership of Upper Newport Bay, most of which has subsequently become a state-operated ecological reserve. The Upper Bay is a marked contrast to the marina complex of the Lower Bay—developments once planned for the Upper Bay. The preservation of Upper Newport Bay is one of Southern California conservationists' best success stories.

The wetland is a premier bird-watching spot. Plovers stand motionless on one leg, great blue herons pick their way carefully across the mudflats, flotillas of ducks patrol the shallows. Out of sight, mollusks, insects, fish and protozoa provide vital links in the complex food chain of the estuary.

This hike follows one-way Back Bay Road which really should be closed to motorized traffic. However, on weekdays, there's rarely much traffic and on weekends, there's seldom more auto traffic than bike traffic. The tideland is fragile; stay on established roads and trails.

Directions to trailhead: Turn inland off Coast Highway onto Jamboree Road, then left on Back Bay Drive. This road follows the margin of the bay. Park along the road.

The Hike: As you walk along the road, notice the various vegetation zones. Eel grass thrives in areas of almost constant submergence, cord grasses at a few feet above mean low tide, salt wort and pickleweed higher on the banks of the estuary. Keep an eye out for three of California's endangered birds: Beldings Savanna Sparrow, the California least tern, the light-footed clapper rail.

Old levees and an occasional trail let you walk out toward the main bodies of water. Also, a trail from the University of California at Irvine runs along the west side of the reserve.

Return the same way.

Huntington Beach Trail (0-5)

Huntington State Beach to Huntington Pier, 2 miles RT
Huntington State Beach to Bolsa Chica State Beach, 6 miles RT
Huntington State Beach to Bolsa Chica Lagoon, 8 miles RT

In the nineteenth century, the country back of Huntington Beach was lush with willows, springs, and peat bogs. A beautiful river rose from where present day Westminister stands. It flowed south, sweeping along the base of Huntington Beach Mesa and entering the sea through Bolsa Bay. In the 1920s, during the citrus boom, great numbers of artesian wells and drainage ditches were dug and the river disappeared.

Before Huntington Beach received its present name, the long shoreline was a popular camping spot. Millions of small clams were washed up on its sands and old timers called it Shell Beach. In 1901 a town was laid out with the name Pacific City, in hopes it would rival Atlantic City. In 1902, Henry E. Huntington, owner of the Pacific Electric Railroad, bought a controlling interest and renamed the city after himself.

This hike takes you along wide sandy Huntington State Beach to Bolsa Chica State Beach and adjacent Bolsa Chica Ecological Reserve.

Directions to trailhead: This is an ideal beach trail to bike 'n hike. A bike path extends the length of Bolsa Chica State Beach to the Santa Ana River south of Huntington Beach. You can leave your bike at Bolsa Chica Beach and hike up to it from Huntington Beach. To reach Huntington State Beach, take the San Diego Freeway (405) to Beach Boulevard. Go west to the park entrance. Mass Transit: OCTD #1 and #35.

The Hike: Walk north along the northernmost mile of the three-mile-long sandy state beach. Pismo clams are found here and the area has seen a strong revival of clamming in recent years. Also keep an eye out for *Donax gouldii*, the small bean clam, that frequents the intertidal zone. The bean clam ends up in many local chowders. The bean clam lives just below the surface of the sand and is often accompanied by the one-inch long hydrozoa *Clytia Bekeri*, which extends above the surface of the beach.

The state beach blends into Huntington City Beach just before the pier. The beach is best known as the site of international surfing competition. eighteen-hundred-foot Huntington Pier was built in 1914. It has a bait-and-tackle shop and is floodlit at night for fishing and surfing.

Beyond the pier is 6-mile-long Bolsa Chica State Beach. Only the northern three miles offer facilities. The southern end has steep cliffs rising between Pacific Coast Highway and the beach. Huntington Beach Mesa or "The Cliffs" is popular with surfers and oil well drillers.

Bolsa Chica Lagoon Loop Trail (0-6)

3 miles RT

Bolsa Chica Wetlands, an 1800-acre tidal basin surrounded by the city of Huntington Beach, is one of Southern California's most valuable oceanfront properties. The somewhat degraded marshland is the scene of a decade-long dispute between Signal Oil Company, the principal landholder, which would like to develop a marina and suburb, and Amigos de Bolsa Chica, who would like to preserve the marsh as a stopover for migratory birds on the Pacific Flyway and as a habitat for endangered species.

For many centuries the wetlands were the bountiful home of Indians until a Mission era land grant gave retiring Spanish soldier Manuel Nieto title to a portion of Bolsa Chica. Although the coastal march proved useless for farming and ranching to Nieto and succeeding owners, the abundant wildlife attracted game hunters from all over Southern California. At the turn of the century, the Bolsa Chica Gun Club, a group of Pasadena businessmen, acquired title to the land and created a hunting preserve. In order to stabilize their duck pond, they dammed off the ocean waters, thus starting the demise of the wetland.

In the 1920s, oil was discovered at Bolsa Chica. Dikes were built, water drained, wells drilled, roads spread across the marsh. In fact, oil production is scheduled to continue through the year 2020.

Portions of the marsh bordering Coast Highway have been restored by the state and are now part of an Ecological Preserve. This loop trail takes you on a tour of the most attractive section. Bring your binoculars. Birding is quite good here.

Directions to trailhead: Bolsa Chica Ecological Reserve is located just opposite the main entrance of Bolsa Chica State Beach on Pacific Coast Highway. From the San Diego Freeway (405) exit on Beach Boulevard, follow it to the beach. Head north on Pacific Coast Highway for three miles to the Reserve entrance.

The Hike: At the trailhead is a sign giving some Bolsa Chica history. Cross the lagoon on the bridge, where other signs offer information about marshland plants and birds. The loop trail soon begins following a levee around the marsh. You'll pass fields of pickleweed and cordgrass, sun-baked mudflats, the remains of oil drilling equipment. Three endangered birds: Savanna sparrow, clapper rail, California least tern, are sometimes seen here.

At the north end of the loop, you may bear right on a closed road to an overlook. As you return, you cross the lagoon on another bridge and return to the parking area on a path parallelling Pacific Coast Highway.

Bolsa Chica Beach

San Mateo Canyon Trail (O-7)

Forest Road 7S02 to Granite Trail Junction
5 miles RT; 100' loss

Forest Road 7S02 to Clark Trail Junction
11 miles RT; 400' loss

Forest Road 7S02 to Camp Pendleton
18 miles RT; 1100' loss

Two hundred-year-old oaks, tangles of ferns, nettles, and wild grape, and the quiet pools of San Mateo Creek make the bottom of San Mateo Canyon a wild and delightful place. This section of the Santa Ana Mountains is steep canyon country, sculpted by seasonal but vigorous streams. San Mateo Creek, a cascading waterway in winter, slows to a gurgle in summer and flows above ground only sporadically in fall. Wilderness designation for this section of the Cleveland National Forest would protect the headwaters and watershed of San Mateo Creek.

San Mateo Canyon takes its name from one of the padres' favorite evangelists and holy men. It's the crown jewel of the Santa Ana Mountains, a relatively untouched wilderness of oaks, potreros and cattail-lined ponds. It's a haven for turtles and rabbits, hawks and quail. Spring brings prolific wildflower displays. The canyon drops from 3,500 feet to the coastal plain at Camp Pendleton.

This day hike or backpack plunges through the southern part of San Mateo Canyon, easily the wildest place in the Santa Ana Mountains. The San Mateo Canyon and other riding and hiking trails in the area have been in use since the turn of the century, Orange County Sierra Clubbers have worked tirelessly on the trail, but it still has rough places. The trail is supposed to stay close to the creek and only slosh through it occasionally. However, brush nearly covers the trail frequently and poison oak drives the susceptible into the creek. You can journey almost as far down canyon as you like in one day. It's 9 miles from Fisherman's Camp to the marine base, with a hundred ideal picnic spots and several fine camp sites along the way.

Directions to the trailhead: From Interstate Highway 15 near Wildomar, exit on Clinton Keith Road, proceeding 7 miles to the Tenaja turnoff. Turn right on Tenaja Road (7S01). Pass enaja Fire Station and continue another 3 miles on 7S01 until it intersects with Forest Road 7S02. If 7S02 isn't washed out, follow it to Fisherman's Canmp.

The Hike: If 7S02 is closed, hike a long mile to Fisherman's Camp. In spring wildflowers line the road. Once, many "fisherman's camps" lay along San Mateo Creek. In the thirties, anglers were attracted by superb fishing for steelhead trout. The San Mateo Canyon Trail was the best route to their favorite fishing holes. Steelhead ran into the forest up San Mateo Creek in 1969 and will again after a heavy rain and runoff.

From Fisherman's Camp you have a choice of routes. San Mateo Canyon Trail (5W05) ascends a hill on the canyon's south side, then soon switchbacks down near the creek. It follows the west side of the creek until the Bluewater Trail Junction (hard to spot) ½ mile away. However, the trail is often overgrown with brush and poison oak. In my view, it's easier and prettier to simply pick your own trail and follow San Mateo Creek downstream.

Whichever route you choose, past the Bluewater Trail Junction, the San Mateo Canyon Trail passes through long narrow meadows, once in a while skipping down to the creek. One and a half miles from Fisherman's Camp, you'll reach Bluewater Canyon and another trail junction. A sign here indicates it's 3 miles up the Granite Trail to Oak Flats. San Mateo Canyon Trail continues 3 miles to the intersection with the Clark Trail. You can picnic under the oaks near the junction and return, or continue down the canyon.

Option: San Mateo Canyon to Clark Trail. Continue down the creek ont the San Mateo Creek Trail. The boulders get bigger, the swimming holes and sunning spots, nice. One flat rock, popular with Sierra Clubbers, has been nicknamed "lunch rock." The trail takes you under ancient oaks and sycamores and along the cattail-lined creek. As you near the Clark Trail, the San Mateo Creek Trail utilizes part of an old mining road. Beyond the Clark Trail junction San Mateo Creek Trail soon peters out and the route down canyon is trailless to Camp Pendleton. The Clark Trail, if you're game, ascends very steeply 1½ miles to Indian Potrero.

Option: San Mateo Canyon to Camp Pendleton. The last three miles of San Mateo Canyon is sometimes wide, sometimes narrow, and always quite beautiful. It's also rugged and requires many creek crossings. At the 9 mile mark, the canyon mouth opens into a broad coastal plain, cooled by Pacific breezes. You'll see a guard station, where you'll need to turn around to steer clear of Camp Pendleton.

Marines will continue across Camp Pendleton and join the CCT on the coast; civilians must return to the trailhead up canyon.

Bear Canyon Trail (0-8)

Ortega Highway to Pigeon Springs; 5.4 miles RT; 700′ gain
Ortega Highway to Sitton Peak; 10.4 miles RT; 1300′ gain

Refreshing Pigeon Springs welcomes hot and dusty hikers to a handsome oak glen. If you hike to the springs early in the morning, before other visitors arrive, you may see a coyote, bobcat, or deer sipping the cool waters. Shutterbugs often hide in the foilage near the spring and await the perfect wildlife photo.

The recently completed Bear Canyon Trail climbs through gentle brush and meadow country, passes Pigeon Springs and arrives at Four Corners, a meeting place of several major hiking trails that tour the southern Santa Anas. One of these trails takes you to Sitton Peak for a fine view. In the future, if the Forest Service constructs connecting trails from Teneja Station and Fisherman's Camp (See San Mateo Canyon Trail), the Bear Canyon Trail will allow hikers to reach San Mateo Canyon from Ortega Highway.

Directions to trailhead: Take the Ortega Highway (74) turnoff from the San Diego Freeway (Interstate 5) at San Juan Capistrano. Drive east 19½ miles to the paved parking area across from Ortega Oaks store. The Bear Canyon Trail starts just west of the store on Ortega Highway.

The Hike: From the signed trailhead on Ortega Highway, the broad, well-graded trail climbs slowly up brushy hillsides. In spring, the slopes are a delight for wildflower lovers. The trail crosses a seasonal creek, which runs through a tiny oak woodland. After a mile, there's a deceptive fork to the left. Ignore it. The trail climbs in, skirts the periphery of a meadow and crests a hot chaparral covered slope. Just before the trail joins Verdugo Truck Trail, enjoy the view down into San Juan Canyon. Turn right (south) on the truck trail and proceed ¾ mile to Pigeon Springs.

Pigeon Springs, which has water about half the year, includes a storage tank and a concrete trough. The springs are located among oaks on the left of the trail. If the bugs aren't biting, this can be a nice place to picnic.

Option: Pigeon Springs to Four Corners, Sitton Peak. Hike down the truck trail another ½ mile and you'll arrive at Four Corners, a convergence of truck trails. The Verdugo Truck Trail pushes straight ahead to an intersection with the Blue Water Trail. To the left is the Blue Water Truck Trail leading down to Fisherman's Camp. To proceed to Sitton Peak, bear right on the Sitton Peak Truck Trail.

Follow the truck trail as it begins to climb and contour around the peak. There are a few trees up on the ridge, but little shade en route. In a mile you'll be at the high point of the Sitton Peak Truck Trail, a saddle perched over San Juan Canyon. The high point is approximately at Forest Service marker W-56. Follow this trail another mile until you reach the southeast face of the peak. Leave the trail here and scramble up the rocky outcroppings of Sitton Peak. On a clear day, there are superb views of the twin peaks of Old Saddleback (Mount Modjeska and Mount Santiago), Mount San Gorgonio and Mount San Jacinto, Catalina and the wide Pacific.

Bell Canyon Loop Trail (0—9)

5 miles RT through Ronald W. Caspers Wilderness Park

Ronald W. Caspers Wilderness Park, located on the western slope of the Santa Ana Mountains, offers the hiker 5500 acres of gentle canyonland to explore. Huge sycamore trees, wildflowers and chaparral line beautiful Bell and San Juan Canyons. The land was part of Starr Viego Ranch until 1974 when it was purchased by the County of Orange. The park honors County Supervisor Caspers who was instrumental in preserving the land.

Deer, rabbits, and coyote thrive here as well as more furtive animals such as foxes, badgers, and bobcats. Birders will want to pick up a free bird check-list at the park entrance and test their skill by identifying the many species found here.

There's a map and guide available to Caspers' trails, but both are quite out-of-date. Instead, rely on your sense of direction as you follow the trails, numbered 1 through 6. The loop trail outlined below, one of many possible loops through the park, offers a sampling of the parks various ecological communities—riparian oak, meadowland, chaparral. This day hike is a fine introduction to the Santa Ana Mountains and brings you ocean and mountain views from atop the park's west ridge.

Directions to trailhead: Exit Interstate 5 in San Juan Capistrano at the State 74 turnoff. Head inland 8 miles on 74 (Ortega Highway) to the park entrance. Park in the day use area near the corral and windmill. There's a day use fee.

The Hike: From the day use area, follow the signed Nature Trail, crossing Bell Creek and looping through a lush oak grove. The Nature Trail returns you to the creek a little farther up canyon. You re-cross the creek and bear left on Trail 3, the Bell Canyon Road Trail. The trail follows nearly level Bell Canyon, which eventually opens up and joins large San Juan Canyon. The trail meanders through meadowland, gold and crinkly in summer and fall, lush and green in spring. Oak copses dot the meadow. Red-tailed hawks roost atop spreading sycamores. You are never very far from Bell Creek, its stream bed and sandy washes.

At the northern end of Trail 3 you can cross Bell Creek and follow Trail 5, paralleling the creek south for a few hundred yards. The trail then climbs through a handsome oak grove up the west slope of Bell Canyon. At the summit ridge on the park's boundary, a fine view is yours. To the north is Saddleback Mountain, the highest peak in Orange County, approximately

5200'. To the west and south you can see the town of San Juan Capistrano, the wide blue Pacific, and on an exceptionally clear day, San Clemente Island.

Return the same way on Trail 5. Back at the junction with Trail 3, you can bear north then east on Trail 1, following it along the park's boundary with the Audubon Reserve, and then heading south on a gradual descent back through Bell Canyon. A short connector trail will return you to Bell Canyon Road (Trail 3) and back to the trailhead.

Wild Cucumber

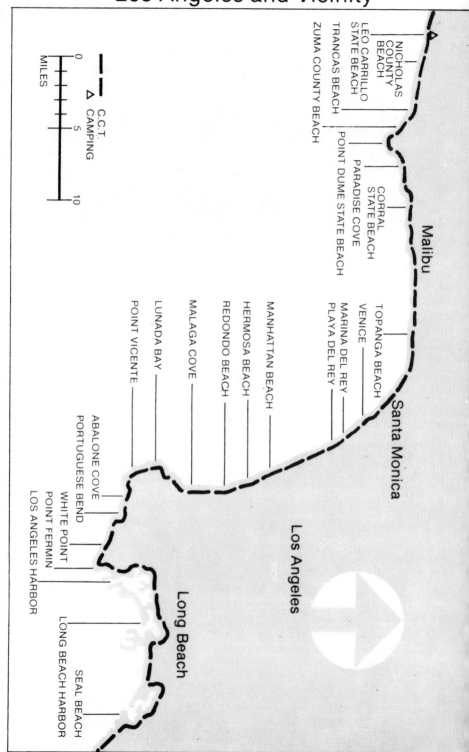

Los Angeles County

Portola Plaque

When the summer sun beats down on the metropolis and smog thickens, half the Southland flees to Los Angeles County's seventy-four miles of coastline. Fifty million visits a years are made to county beaches, although, as CCT hikers soon discover, most cluster blanket-to-blanket on the same beaches, leaving less accessible areas to those willing to hike. The typical mass-use L.A. beach includes acres of hot sand, waves ranging from the gentle to the inspired, a lifeguard every few hundred feet, and a boardwalk full of roller skaters, restaurants and raft rental establishments. Before dawn, huge mechanized sandrakes scoop up trash, doing a good job of picking up after sloppy beachgoers.

At the south end of the county is long, break-water protected Long Beach and the Long Beach-San Pedro Harbor complex, which offers little to the outdoors-minded except boats bound for Catalina Island. Located 22 miles offshore, Catalina, with its steep brush and cactus-covered ridges, clear waters, and beautiful coves, is a delight for hikers. Most of the island, except for the resort town of Avalon, is rural, the domain of buffalo, boar, and backpackers.

Rocky Palos Verdes Peninsula has been the death of many ships. The peninsula's 15 miles of reefs, tidepools, coves and crescent beaches, located far below exclusive cliff-top homes, will surprise those energetic enough to hike it. Marineland, with extensive aquariums housing marine life from all over the world, tanks full of whales, seals and dolphins, and performing animal shows, is located on the peninsula.

A nearly uninterrupted stretch of sandy beach extends from the north end of Palos Verdes Peninsula, from Redondo Beach to Malibu. Each city and beach along this strand has a unique personality: Venice with its canals and zany boardwalk, Hermosa with its surfer population, Marina Del Rey with the largest man-made pleasure craft harbor in the world.

The Santa Monica Mountains are the only relatively undeveloped mountain range in the U.S. that bisects a major metropolitan area. They stretch from Griffith Park in the heart of L.A. to Point Mugu, fifty miles away. The Santa Monica Mountains National Recreation Area protects a patchwork of state and county parks. The network of trails through the Santa Monicas is a rich pastiche of nature walks, scenic overlooks, fire roads and horse trails leading through diverse ecosystems: native tall grass prairies, yucca-covered hills, and springs surrounded by lush ferns. Hikers

CCT at a glance

TERRAIN: Sand beach ranging from wide (Santa Monica Beach) to narrow (Malibu Colony). 15 miles of rocky shoreline with tidal bores on Palos Verdes Peninsula.

OBSTRUCTIONS: Mammoth Long Beach-San Pedro Harbor can be walked but is best skipped unless your interest is tuna canning factories and petroleum. Rocky Palos Verdes Peninsula is passable at low tide. Marina Del Rey is but a minor interruption. Residences crowd the surf in parts of Malibu. Watch the tides.

TRANSPORTATION: Amtrak stops in L.A. Plenty of bus service. Southern California Rapid Transit District RTD 443-1307, Santa Monica Municipal Bus Lines, 451-5445, Long Beach public Transit 591-2301.

CAMPGROUNDS/ACCOMMODATIONS: Of the 15 coastal counties in California, Los Angeles with the largest population, has the least number of campsites. In fact, it has exactly one—Leo Carrillo State Beach at the extreme north end of the county. Only one hostel operates all year and it's not very close to the coast.

    Leo Carrillo State Beach
    Westchester YMCA (open June 1 to Sept. 15 only)
    Los Angeles International Hostel

are hoping that the Backbone Trail, to run from Will Rogers Park to Point Mugu State Park will soon be completed and become an important link of the CCT.

The northern part of the county's coastline, beginning about Malibu, is decidedly different from the south. The Santa Monica Mountains veer toward the coast, creating a series of bluffs, rocky points, coves, and sandy beaches. This coastline is the most rural in the county, although there are occasional clusters of expensive homes built, it seems, right over the surfline. The Coastal Commission has been improving access to a number of secluded (and in some cases unnamed!) beaches that are far from the maddening crowd.

CCT: LOS ANGELES COUNTY

After crossing the San Gabriel River into Los Angeles County, the hiker will improvise a route through the Long Beach Marina along Alamitos Bay to Marina Drive. Bearing left here, CCT continues through the marina ¼ mile to Westminster Avenue. Go left on Westminster over the Long Beach Marine Stadium Bridge, then descend immediately to your right to Marine Park. The park is composed of a narrow sandy beach and grassy area. CCT follows the beach northwest over the sand. Offshore to your right is the mile-long rectangular body of water known as Marine Stadium. It was built in the 1920s and was the site of the 1932 Olympics rowing competition. Super-charged drag boats reaching 200 mph have raced here and the stadium hosts regattas and kayak races.

Since there are few hills offering panoramic overviews, hikers may not realize they are walking on an island called Naples (three islands, actually, separated by narrow canals). Naples Land Company acquired the marshy site here and planned a cottage community connected with bridges and gondolas. Sea walls were built to raise the land to a height of six feet, canals dredged and "Naples of America" came into being. Lots were offered for sale in 1906, but the real estate market was not favorable. The land passed into the hands of Henry E. Huntington and the development was not completed until 1923. Prospective buyers arrived at Naples via Pacific Electric railway and took a side-wheeler boat trip around the bay, all the while receiving a sales talk. The Naples canals were ruined in the Long Beach Earthquake of 1933, but rebuilt with the help of the WPA during the Great Depression.

CCT leaves Naples Island, returns to the mainland via the Appian Way Bridge, then makes a left to Bay Shore Avenue. Hikers follow the avenue around the horseshoe of Alamitos Bay to Belmont Shores (Long Beach)

Beach. A mile walk on sandy beach brings you to Belmont Pier, a 1300-foot fishing pier.

Stretching ahead is Long Beach, well named, for it is both long and wide. The town was first named Willmore City for W.E. Willmore, who developed it in 1880 and colonized it with emigrants from Kansas City. Few lots sold. In 1887, the Long Beach Land and Water Company took over, giving the community its name.

CCT follows Long Beach City Beach for nearly 5 miles beneath bluffs. The 100-yard wide beach offers some of the finest and most gentle ocean swimming in Southern California. An ocean breakwater protects the shore from all but the most southerly swells.

As CCT nears an area of construction, you'll join Ocean Boulevard, heading due west toward downtown. Alamitos Street heads over the Queensbay Bridge to the *Queen Mary*. Pedestrians are not permitted on the bridge. The *Queen Mary*, the largest passenger ship ever built, over 1,000 feet long with a crew of 1200, was launched in 1934. The British luxury liner served as a troop ship during World War II. It was retired in 1964 and is now a tourist attraction and hotel.

Older hikers will remember the 400 block of Ocean Boulevard as the site of The Pike, established July 4, 1902, and for over six decades the most popular amusement park on the West Coast. Attractions included a bath house known as The Plunge, the Walk of a Thousand Lights, three ball-rooms and the Cyclone Racer Roller Coaster.

Ahead of you is the Los Angeles River. The trailhead for the Lario Trail is at the Golden Avenue Boat Launch at the end of Golden Shore Boulevard. Lario (Los Angeles River-Rio Hondo River) Trail is a 25-mile bikeway stretching through the heart of the L.A. basin from Long Beach to Montebello.

In time, CCT may negotiate the mammoth Long Beach-Los Angeles harbor complex, but for the present and immediate future, it holds interest for only the most industrially-minded hikers and those wishing to make tuna cannery tours. Heavy traffic and two bridges prohibiting pedestrians, the Desmond Bridge and the Vincent Thomas Bridge, discourage walking from Long Beach to San Pedro. The harbor complex is best viewed from a bus window.

Call RTD (443-1307) or Long Beach Public Transit (591-2301) and improvise a route north to Cabrillo Beach.

From San Diego to San Francisco, California has no natural harbor. The huge federally built breakwater provides the protection that Nature didn't. Long Beach Harbor is connected with the Los Angeles Harbor by Cerritos Channel, spanned by a giant lift bridge, which carries traffic to Terminal Island. Extensive cargo-handling facilities are here. It's hard to believe that almost none of this harbor existed before 1940.

Long Beach

L.A. Harbor, with 28 miles of waterfront is one of the largest artificial harbors in the world. America's largest fishing fleet and a huge canning center is located here. In fact, large commercial catches of halibut, yellow tail, and sea bass were made right in the harbor until about 1950. Water quality was deliberately allowed to deteriorate to protect wood structures from ship worms and other boring organisms. The reproductive habitat of the fish was, of course, eliminated.

As long as you're riding buses, consider busing to the L.A. International Hostel. Considering no campgrounds or low-priced accommodations are located along the CCT in L.A. County, you may want to avail yourself of the hostel's hospitality: 1502 Palos Verdes Drive North, Harbor City 90710, (213) 831-8109. It's a mile walk from Cabrillo Beach to the bus stops at 36th and 13th Streets. Take RTD bus #849, which will drop you off on Palos Verdes Drive.

Cabrillo Beach is really two beaches, a still water beach inside the breakwater and a surf beach on the ocean. At Cabrillo Beach is a statue to Juan Rodriguez Cabrillo, the first European to set foot in California. In 1542 he named the slight coastline indentation here, "The Bay of Smokes." Historians guess the Indians were setting the grass afire for their periodic rabbit hunts; hence the name.

Across from Cabrillo Beach is the Cabrillo Marine Museum with exhibits of marine invertebrates and salt water aquariums, sculptures and murals.

Departing Cabrillo Beach, CCT rounds Point Fermin; passage is possible only at low tide. Young author-adventurer Richard Henry Dana threw cowhides down the cliffs here, though his name was later applied to a point in Orange County, site of more hide throwing. Dana found the Palos Verdes headlands stark and oppressive; perhaps it's difficult to appreciate nature when busily engaged in hide loading.

Tidepools at the point are protected by Point Fermin Marine Life Refuge and are full of star fish, sea urchins, anemones, octupus, and hermit crabs. On the bluffs is a Victorian style light house, built from materials shipped around Cape Horn. Daring hang gliders soar overhead.

CCT follows the rocky shoreline to White's Point. Offshore dedicated volunteers have built an underwater nature trail for divers and snorkelers. Around the point to the west CCT reaches Royal Palms Beach State Park. The rocky cove here flourished during the Roaring 20s as a hot spring resort and health spa. A few of the resort's foundations and walls remain, as well as wonderfully overgrown terraced gardens and fine palm trees.

Beyond, stretch miles of rocky shore. There's virtually no beach access here, so you'll probably have the rocks to yourself. Portuguese Bend, three miles up-coast from Royal Palms is one of the most geologically active sites in Southern California and evidence of slumping is obvious. Earth movements in 1956 destroyed 100 homes. The rate of land movement reached slightly more than an inch a day (!) in 1957, but has slowed since to a few inches a year.

Shore whaling was carried on along California in the 1850s and a string of whaling stations was established on the coast from Baja to Half Moon Bay. One station was Portuguese Bend, named for its many Portugese whalers.

Up-coast from Portuguese Bend, just beyond the earth-slide area, is Inspiration Point. The shoreline is impassable right at the point due to a spectacular tidal bore. Follow the unmarked trail to the bluffs. A short trail takes hikers to the incredible view from the top of the point and to a most inspiring picnic spot. Santa Catalina Island may be seen and, on an exceptionally clear day, beyond the western part of Catalina, Santa Barbara Island can be distinguished; a total distance of more than 50 miles.

> "In Palos Verdes one has the impression of entering a paradise designed by the Spanish for the annointed of Heaven." —Louis Bromfield, *Vogue,* 1930

CCT descends from Inspiration Point via a good, but steep trail to the brief crescent-shaped beach, follows the beach, then ascends another goat trail up the bluffs at Portuguese Point. There are tidal bore caves at the point; hence the second detour. Portuguese Point is equally inspiring. You follow a trail atop the headlands, then descend another goat trail to Abalone Cove.

Abalone Cove's warm shallow water offers some of the finest snorkeling in L.A. County. Divers explore rocky outcroppings encrusted with urchins and anemones and a thriving kelp bed. Landlubbers get great views of the peninsula's terrace formations which flow in long graceful curves from rocky cliffs to the buff-colored bluffs.

As you hike up-coast from Abalone Cove, you may see natural tar seepage, prevalent on the peninsula. Indians caulked the seams of their boats with the asphaltum and paddled their watertight craft to Catalina Island. You'll soon pass the private facilities of the Abalone Cove Club, west of Portuguese Point. Indian artifacts including shell beads, mortars and even skeletons have been found here.

More rock-hopping brings you past the Marineland Pier to Long Point, the southwestern extremity of the peninsula, only 17 nautical miles from Catalina. The point is occupied by Marineland. Opened in 1954, Marineland features massive tanks, trained whales, performing porpoises and sea lions, and hundreds of variety of ocean life from leopard sharks to giant turtles. It's well worth a visit.

Around the wide point, hikers reach Point Vincente Fishing Access, a rocky area favored by surf fishermen and divers. On the cliffs above perches Point Vincente Lighthouse, open to visitors Tuesday and Thursday afternoons. The stretch of shore beyond the lighthouse has been vigorously scalloped by thousands of years of relentless surf. Boulders are big, passage difficult.

As you near Resort Point, where fishermen try their luck, you'll see numerous stacks, remnants of former cliffs not yet dissolved by the surf. CCT proceeds under the almost perpendicular cliffs that follow horseshoe-shaped Lunada Bay. At Rocky (also called Palos Verdes) Point, the remains of the rusting Greek freighter *Dominator* lie imprisoned on the rocks. As you round the point, Santa Monica Bay and the Santa Monica Mountains are visible as gray silhouettes.

Bluff Cove, a popular surfing spot, gives way to anvil-shaped Flat Rock Point with a superb tidepool area. Around this last point is sandy Malaga Cove, the only sandy beach on the Palos Verdes Peninsula—with the exception of Cabrillo Beach adjoining the Los Angeles harbor. As Malaga Cove straightens out, the steep-walled terraces grow shorter and before you stretch miles and miles of sandy South Bay beaches.

As the cliffs recede, CCT crosses Torrance Beach, popular with surfers and swimmers. Marching over the white sands, hikers soon arrive at wide Redondo State Beach. The town of Redondo Beach was founded in 1892; its name is derived from nearby Rancho Sausal Redondo (Round Willow Grove).

Ahead, toward the north end of the state beach, is King Harbor. In the 1890s, it appeared that this harbor might become the great port of Los Angeles. Great vessels anchored off shore, goods from Asia and the Pacific Islands were unloaded at the wharves, and railroad tracks ran from

Redondo docks to downtown Los Angeles. Congress considered Santa Moncia, San Pedro and Redondo for federal funding to build a sea port. San Pedro Harbor got the dredging and improvement monies and Redondo became a resort. An amusement park and salt water plunge hosted generations of Southern Californians. During Prohibition, "entertainment" boats anchored off Redondo shores, just outside the 3-mile limit.

King Harbor today bears little resemblance to Redondo's busy little port of the 1890s. The harbor area is dominated by three piers: Monstad Pier, a 300-foot privately owned fishing pier, Municipal Pier, also known as Horseshoe Pier, full of restaurants and shops, and another private pier, Sportsfishing Pier.

CCT improvises a route through King Harbor. Hikers may dawdle along the walkways around the boat basins or follow the South Bay Bicycle Path to the east. Note Seaside Lagoon, a warm 2.5 acre warm saltwater swimming hole with a sandy bottom, located between Basins 2 and 3.

At the north end of Redondo Beach is the site of an old salt lake, from which Indians obtained salt. In the 1850s, a company began using artificial and solar evaporators to manufacture salt. Peak year was 1879, when 450 tons were produced.

Hermosa Beach Municipal Pier, a 1320-foot fishing pier at the south end of Hermosa City Beach is soon passed. Hermosa is Spanish for beautiful; the name was supplied by a real estate developer. Hermosa's two-mile long, wide sandy beach is paralleled by a scenic walkway established in 1908 known as The Strand. It hosts every beach culture activity imaginable. In the 1950s crowds of beatniks hung out in Hermosa Beach coffee shops and bookstores, listening to jazz and poetry read to music.

Manhattan State Beach, soon crossed by CCT, is yet another wide sandy beach and also features The Strand. Manhattan Pier a fishing pier, is located at the south end of the beach. In the early twentieth century, a wave-powered generator was located here, which produced electricity to light The Strand. The pier suffered heavy storm damage in the 30s and 40s and was rather run-down and seedy, judging from the description in one of Raymond Chandler's short stories.

A mile past the pier, you'll pass low dunes which served as a "desert" location for many silent movies. Inland squats a massive power plant and "the plumber's nightmare"—the Hyperion Waste Treatment Plant. As the story goes, Aldous Huxley and Thomas Mann were hiking along this beach one day, discussing Shakespeare and brave new world and other lofty subjects, when Huxley noticed what appeared to be thousands of wriggling, pale-colored worms. On closer inspection, the worms turned out to be condoms. At first Huxley suspected typical Southern California orgiastic excess, but soon realized that the Hyperion Sewage Plant was discharging into the sea and the love aids had washed in with the tide. Huxley's beach walk and discovery led him to some profound speculation about man, civilization and sewage and he published a fine essay entitled "Hyperion to a Satyr."

After passing an ugly oil pier, you'll walk along Dockweiller Beach, which is wide, clean, sandy, and mostly pleasant—considering it's backed by L.A. International Airport. Dockweiller, usually referred to as Playa del Rey, extends to Marina Del Rey harbor.

When you reach the Ballona Creek Channel, cross the Balboa Creek Bridge to the harbor jetty and follow the bikeway atop the jetty inland. On your left sailboats tack in and out of the harbor entrance. Occasionally a skipper goofs and sails his craft into the Ballona Creek Channel, inevitably getting marooned on the rocks.

On your right is the remnant of a once extensive marshy area that extended from Playa del Rey bluffs to Venice. Housing developments, Army Corps of Engineer flood control projects, and Marina del Rey all but obliterated the marsh. Still, some pickleweed wetlands and mudflats remain, providing an important rest stop for birds along the Pacific Flyway. Ballona Creek is the winter home for many species of ducks, grebes, and loons.

Development pressures leave the future of L.A. County's only remaining wetland very much in doubt.

Past the UCLA rowing crew headquarters and some apartment buildings, the bikeway you've been following forks. CCT goes left. By continuing right on the Ballona Creek bikeway one half mile to Lincoln Boulevard, following Lincoln to Sepulveda, and walking up Sepulveda one mile, you'll reach the Westchester YMCA Hostel at 80th & Sepulveda. Or RTD #871 to Manchester and Sepulveda Blvd; walk four blocks north on Sepulveda to 80th Street. From L.A. Airport, take RTD #88. The hostel has 35 beds. Open June 1-September 15 (213) 776-0922.

CCT follows a bike route and walkway past Fisherman's Village and various boat basins and crosses Admiralty Way. Here you follow the landscaped bikeway along a bird refuge to Washington Street, then head west on Washington toward Venice Beach. On the west side of Washington Street, note one of the famed canals. Venice-By-The Sea, was the hallucination of Abbot Kinney, the manufacturer of Sweet Caporeal Cigarettes. He was, as they say, a rich man with artistic impulses. He proceeded to drain the swampland here and build cottages connected by canals. He imported gondolas and gondoliers from that other Venice across the Atlantic and built a Chataqua hall for the education of the citizenry. Then as now, however, the tawdry amusements of the boardwalk proved to be more popular than art and education. Today, the canals look a bit worse for wear. The system is choked with weeds and not much thrives there except ducks.

CCT returns to the beach at the Venice Fishing Pier. Marina Peninsula (Venice Beach) extends downcoast one mile to the Marina Del Rey Channel entrance. CCT heads up-coast along wide, sandy Venice City Beach. Venice is one of the county's oldest city beaches; lifeguard service has been provided since the 1920s. The quiet of this beach is a marked contract to the hubbub of Ocean Front Walk paralleling the length of it. Almost every facet of zany California beach culture can be found on the boardwalk—jugglers, mimes, a disco roller rink, and more sidewalk cafes, junkfood and healthfood eateries than one could sample in a year of Sundays.

CCT crosses Santa Monica Beach next, an extremely wide beach, bisected by Santa Monica Pier. The southern section has more tanners per square yard than does the north. This beach is perhaps no longer the glamour spot it was of old, but 5,000 parking spaces attest to the fact that a lot of people find it quite attractive.

Rising 40-60 feet above Santa Monica Beach are the Palisades cliffs. The alluvial sands and gravels are in nearly horizontal layers and subject to severe erosion. The alluvium was deposited here by streams cascading down from the Santa Monica Mountains. Ages ago, Pacific waves found these deposits easy going and cut back extensively at the terraces, forming the

Palisades. Palisades Park atop the cliffs at the north end of Santa Monica Beach is popular with joggers, tourists and sunset promenaders. Santa Monica Municipal Pier features the Playland Amusement Arcade, a relic from palmier days, and a wonderful carousel with elaborately carved prancing horses.

CCT continues past the pier on wide sandy beach. The Palisades rise higher and higher, to more than one hundred feet above the beach and highway. Extraordinary (or comical depending on your point of view) attempts to stabilize the cliffs may be seen along this section of the CCT. Three more miles of sandy beach follow as CCT crosses Will Rogers State Beach, named for the famed cowboy and radio personality of the 1930s. There's usually not much surf here, but it has its moments and the beach is well used by swimmers and tanners.

As CCT proceeds west of Sunset Boulevard, the beach narrows and the shoreline grows sporadically rocky. Topanga State Beach is a mixture of sand and rock. The mouth of Topanga Creek is a gathering place for surfers. Sedimentary sea cliffs rise close to shore. In one more mile CCT crosses another narrow state beach, Las Tunas, held together by a series of submerged groins.

You'll hike past the neighborhood beaches of Las Flores, La Costa and Big Rock. Beyond Big Rock is the infamous Zonker Harris Accessway, a focal point of the ongoing debate between the California Coastal Commission, which is determined to provide access to the coast, and some Malibu residents who would prefer the public stay out. The original sign read "Zonker Harris Memorial Beach," honoring a character from the Doonesbury comic strip whose primary goal in life is to acquire the perfect tan.

Just up-coast from Zonker's beach is the Malibu Pier, built in 1903. The 700-foot pier was restored in 1946 and is now owned by the state.

CCT rounds Malibu Point and arrives in Surfrider Beach (Malibu Lagoon State Beach). When natives say "Malibu Beach" this is what they mean: the site of beach blanket movies and Beach Boy songs. Surfrider is a mixture of sand and stone and lots of loose surfboards are flying about. Just inland, Malibu Lagoon hosts many different kinds of waterfowl, both resident and migratory. The beach is rock cobble on the ocean side of the lagoon. To the landward side of the lagoon stretches the alluvial fill flatland deposited by Malibu Creek. The town of Malibu is situated here.

Around Malibu Point, you begin walking across the narrow and sandy beach lined by the exclusive Malibu Colony residences, home to many a movie star. Toward the west end of The Colony, the beach narrows considerably and houses are built on stilts, with the waves pounding beneath them. Depending on the tide, hikers may occasionally be forced to walk *under* the houses.

As you walk along Malibu's beaches, rejoice that you do not see State Highway 60, the Malibu Freeway. In the 1960s a plan was hatched to build a causeway along Malibu Beach, supported on pilings offshore. A breakwater would have converted the open shore into a bay shore. The wonderful pounding surf would have been reduced to that of a lake; the beach biota completely destroyed.

The beach grows wider and more public at Corral State Beach, located at the mouths of Corral and Solstice Canyons. In the 1960s, landslide-prone and seismically active Corral Canyon was the proposed site of the Malibu Nuclear Power Plant.

CCT pushes on around Latigo Point to Escondido Beach, a small sandy beach, deposited by Escondido Creek. Local divers enjoy the clear waters here. A mile west of Escondido is Paradise Cove Beach, accessible only to hikers and those willing to pay. There's a private fishing pier here and the beautiful beach is a favorite locale of movie-makers.

CCT rounds a minor coastline bulge to Dume Cove. A goat trail is taken, ascending to the top of the Point Dume headlands. A number of trails allow exploration of the point. In winter, Point Dume is an ideal place to watch for migratory gray whales who cruise quite close to shore here.

Pt. Dume is a geologist's delight; a combination of sedimentary, sandstone and mudstone, volcanics and shales. It's believed the point captures sand on its up-current side, thus enhancing Zuma Beach and making it one of the finest sand strands on the California coast. After you've taken in the panoramic view from the point, descend the sand cliffs on the goat trail to Westward Beach, a part of Zuma County Beach. Zuma is L.A. County's largest beach. The surf is rambunctious here for swimmers and surfers.

CCT travels up-coast for four beautiful miles across sandy Zuma Beach; the shoreline then begins narrowing again. A series of minor rocky points break up the sandy shoreline. These rocky projections stabilize the beach, but make passage difficult at high tide. The beaches here are also stabilized by the extensive kelp beds offshore. Highly eroded cliffs back narrow Nicholas Canyon County Beach.

Beyond the county beach, CCT crosses into Leo Carrillo State Beach. The park is named for TV actor Leo Carrillo, famous for his role as Pancho, the Cisco Kid's sidekick. There's a walk-in campground for hikers, a vehicle-oriented campground, plus 25 tent sites at the Beach campground. Past the campgrounds CCT hikers scramble around rocky Sequit Point, where there's a series of pleasant caves and coves. It's a good place for tidepool observation. CCT reaches Leo Carrillo Beach North, an undeveloped part of the state beach, and crosses into Ventura County at appropriately named County Line Beach.

Malibu Pier

Meet Catalina Trail (LA-1)

Black Jack Junction to Little Harbor; 8 miles one way
Little Harbor to Two Harbors; 7 miles one way

Catalina Island's terrain is rugged and bold, characterized by abrupt ridges and V-shaped canyons. Many of the mountaintops are rounded, however, and the western end of the island has a small coastal plain. Most of the island is grassland and brush, dotted with cactus and seasonal wildflowers. Only a few native willows and oaks and imported palms attain a size that warrants calling them trees. Bison, deer, boar and rabbits roam the savannahs.

This weekend backpacking trip is a good introduction to the island; it samples a variety of terrain and both sides of the island, west and east, developed and wild. Transportation logistics are a bit complex, but the trails are easy to follow. If you're planning to hike this as a weekend jaunt, it's advisable to take the Friday afternoon ferry and spend the night at the Bird Park Campground, a mile's walk up the hill behind Avalon. From Avalon, you'll be traveling to the trailhead via the Catalina Island Interior Shuttle Bus, which departs from 213 Catalina Avenue (the island's "Main Street") Phone (213) 510-0840 (Avalon) or 510-0303 (Two Harbors). For a few dollars per person, the shuttle bus will drop you off a few miles south of the airport at Black Jack Junction.

See appendix material on Los Angeles County for information on ferry service from the mainland to Avalon. *Caution:* You must tell the ferry company that you will be returning to the mainland from Two Harbors rather than Avalon. They will be happy to accommodate you, but you must tell them. No additional charge.

Catalina Buffalo

The Hike: At signed Black Jack Junction, there's a fire phone and good views of the precipitous west ridges. The trail, a rough fire road, ascends for 1 mile over brush and cactus covered slopes. You'll pass the fenced, but open, shaft of Black Jack Mine. On your left a road appears that leads up to Black Jack Mountain (elevation 2,006). Continue past this junction.

Ahead is picnic ramada with a large sunshade and nearby a signed junction. You may descend to Black Jack Camp, which is operated by L.A. County. Tables, shade, water. We bear right on the signed Cottonwood Black Jack Trail. A second junction soon appears. Continue straight down hill. The other trail ascends to Mt. Orizaba.

The trail descends steeply through a canyon, whose steep walls are a mixture of chaparral and meadowland and are favored by a large herd of wild goats. At the bottom of the canyon pass through three gates of a private ranch. (Close all gates; don't let the horses out.) The trail reaches the main road connecting Little Harbor with Airport-in-the-Sky. You may bear left at this junction and follow the winding road 3½ miles to Little Harbor. For a more scenic route of about the same distance, turn right on the road. Hike about 200 yards to the end of the ranch fence line, then bear left, struggling cross-country briefly through spiney brush and past a small junkyard and intersect a ranch road. This dirt road follows the periphery of the fence line on the east side of the ranch to the top of a canyon. You bear left again, still along the fence line. You ascend and then descend, staying atop this sharp shadeless ridge above pretty Big Springs Canyon. When you begin descending toward the sea, you'll spot Little Harbor.

Two miles of ridge walking brings you to an unsigned junction with Little Harbor Road. The campground is a short distance off this road. It has shaded picnic tables, running water, and a fine sandy beach.

Leave camp on Little Harbor Road, ascending higher and higher into Little Springs Canyon. Buffalo graze both sides of this canyon and two reservoirs have been developed for the animals. At an unsigned junction a mile past Lower Buffalo Reservoir, bear left on Banning House Road. (Little Harbor road continues north, then west to Two Harbors, if you prefer to stick to the road.) Banning House Trail ascends very steeply up a canyon roamed by wild boar. At the windswept head of the canyon, you are rewarded with superb views of the twin harbors of Catalina Harbor and Isthmus Cove and can see both the eastern and western shores of the island. A steep northeasterly descent brings you to the outskirts of Two Harbors. You'll have no trouble improvising a route past ranchettes and private clubs down to the ferry landing.

Royal Palms Beach

Cabrillo Beach Trail (LA-2)

Cabrillo Beach to White's Point; 3 miles RT

All but forgotten today, the rocky cove just downcoast from White's Point once flourished as a Roaring '20s health spa and resort. All that remains today are the sea-battered ruins of a large circular fountain, curious cement foundations, and lush overgrown tropical gardens. White's Point was originally settled at the turn of the century by immigrant Japanese fishermen, who harvested the bountiful abalone from the waters off Palos Verdes Peninsula. Tons of abalone were shipped to the Far East and consumed locally at L.A.'s Little Tokyo. In a few years the abalone was depleted, but an even greater resource was discovered at White's Point—sulphur springs.

In 1915 construction of a spa began. Eventually a large hotel was built at water's edge, palm gardens and a golf course decorated the cliffs above. The sulphur baths were especially popular with the Japanese population of Southern California.

The spa boomed in the Twenties, but the 1933 earthquake closed the springs. The cove became part of Fort McArthur during World War II, the Japanese-American settlers were incarcerated in internment camps, and the resort was soon overwhelmed by crumbling cliffs and the powerful sea. Today, historic societies are seeking California Historic Landmark status for White's Point.

This hike begins at Cabrillo Beach, the only real sand beach for miles to the north and south, passes Cabrillo Marine Museum, visits Point Fermin Marine Life Refuge, and ends up at historic White's Point. The new Cabrillo Marine Museum is well worth a visit. It has marine displays, aquariums with live fish, good shell collections. The museum sponsors tidepool walks, grunion watches, and is a coordinating point for whale watching cruises.

Directions to trailhead: From the Harbor Freeway (State Highway 11) take the Gaffey Street exit and follow Gaffey seaward to 22nd Street; turn left. Continue to Stephen White Drive, turning left into the park. Cabrillo Marine Museum is open 9 to 5 daily. Mass Transit: RTD routes 810, 849.

The Hike: Walk up sandy Cabrillo Beach, which has a monopoly on running grunion, since the sand-seeking fish have few other spawning options along rocky Palos Verdes Peninsula. After passing the San Pedro breakwater and Cabrillo fishing pier, you'll begin hiking around Point Fermin. This is a splendid tidepool area. Keep a look out for sea urchins, octopus, and starfish. On the bluffs above is a Victorian style light house, built from materials shipped around Cape Horn. It was lighted until shortly after Pearl Harbor, when it became an observation room.

The route is rocky as you boulderhop along the base of Point Fermin cliffs. Soon you'll see a palm garden with fire pits. Royal Palms Hotel was situated here until overcome by the sea. Storm twisted palms and overgrown terraced gardens are a reminder of flush times gone by. Royal Palms is a State Beach now, popular with surfers. Ahead at White's Point is the remains of a large fountain. Beyond stretch the rugged cliffs and cobblestone shores of Palos Verdes Peninsula. Return the same way or if you have the time, hike on. The difficult terrain will ensure that few follow in your footsteps.

Palos Verdes Peninsula Trail (LA-3)

Malaga Cove to Rocky Point; 5 miles RT
Malaga Cove to Point Vincente Lighthouse; 10 miles RT

Palos Verdes Peninsula is famous for its rocky cliffs, which rise from 50 feet to 300 feet above the ocean and for its thirteen wave-cut terraces. These terraces, or platforms, resulted from a combination of uplift and sea-level fluctuations caused by the formation and melting of glaciers. Today the waves, as they have for so many thousands of years, are actively eroding the shoreline, cutting yet another terrace into the land.

While enjoying this hike, you'll pass many beautiful coves, where whaling ships once anchored and delivered their cargo of whale oil. Large iron kettles, used to boil the whale blubber, have been found in sea cliff caves. Indians, Spanish rancheros and Yankee smugglers have added to the peninsula's romantic history. Modern times have brought white-stuccoed, red-tiled mansions to the peninsula bluffs, but the beach remains almost pristine. Offshore, divers explore the rocky bottoms for abalone and shellfish. On shore, hikers enjoy the wave-scalloped bluffs and splendid tidepools.

Hiking this shoreline is like walking over a surface of broken bowling balls. The route is rocky and progress slow, but that gives you more time to look down at the tidepools and up at the magnificent bluffs.

Directions to trailhead: To reach Malaga Cove, take Pacific Coast Highway 1 to Palos Verdes Boulevard. Bear right on Palos Verdes Drive. As you near Malaga Cove Plaza, turn right at the first stop sign (Via Corta). Make a right on Via Arroyo, then another right into the parking lot behind the Malaga Cove School. The trailhead is on the ocean side of the parking area, where a wide path descends the bluffs to Malaga Cove.

The Hike: From the Malaga Cove School parking lot, descend the wide path to the beach. A sign indicates you're entering a seashore reserve and asks you to treat tidepool residents with respect. To the north are sand beaches for sedentary sun worshippers. Active rockhoppers clamber to the south. At several places along this hike you'll notice that the great terraces are interrupted by steep-walled canyons. The first of these canyon incisions can be observed at Malaga Cove, where Malaga Canyon slices through the north slopes of the Palos Verdes Hills, then cuts west to end at the cove.

The coastline curves out to sea in a southwesterly direction and popular Flatrock Point comes into view. Jade-colored waters swirl around this anvil-shaped point creating the best tidepool area along this section of coast. Above the point, the cliffs soar to 300 feet. Cloaked in morning fog, the rocky seascape here is reminiscent of Big Sur.

Rounding Flatrock Point, you pick your way among the rocks, seaweed and the flotsam and jetsam of civilization to Bluff Cove, where sparkling combers explode against the rocks and douse the unwary with their tangy spray. A glance over your right shoulder brings a view of Santa Monica Bay, the Santa Monica Mountains in gray silhouette and on the far horizon, the Channel Islands.

A mile beyond Bluff Cove, Rocky (also called Palos Verdes) Point juts out like a ship's prow. Caught fast on the rocks at the base of the point is the rusting exoskeleton of the Greek freighter *Dominator,* a victim of the treacherous reef surrounding the peninsula.

Option: To Point Vincente Lighthouse. Around Rocky Point is Lunada Bay, a good place to observe terrace surfaces. From here you'll walk under the almost perpendicular cliffs that follow horseshoe-shaped Lunada Bay. Shortly you'll round Resort Point, where fishermen try their luck. As the coastline curves south, Catalina can often be seen glowing on the horizon. Along this stretch of shoreline, numerous stacks, remnants of former cliffs not yet dissolved by the surf, can be seen.

The stretch of coast before the lighthouse has been vigorously scalloped by thousands of years of relentless surf. You'll have to boulderhop the last mile to Point Vincente. The lighthouse has worked its beacon over the dark waters since 1926. Lighthouses seem to fall into two categories; those that welcome and those that warn. Point Vincente obviously belongs to the latter category.

Passage is sometimes impossible around the lighthouse point at high tide. If the way is passable, another ½ mile of walking brings you to an official access (or departure) point at Long Point.

Return the way you came, walk up to the bluffs, at Long Point, or take the optional route below.

Option: Bluff Trail to Point Vincente. From Resort Point to Point Vincente, there are a number of "goat trails" that allow the beachwalker to scramble to the top of the bluffs. No residences exist on this section of the bluffs, only a jungle of fennel and a trail to Point Vincente. The trail, offering far-reaching views of the peninsula and Pacific, winds through licorice-smelling headlands to the lighthouse.

Backbone Trail #1 (LA-4)
(Topanga State Park)

Eagle Rock/Eagle Springs Loop; 6.5 miles RT; 800' gain
Eagle Rock to Will Rogers State Park; 10.4 miles one way; 1800' loss from Fire Road 30

For many years hikers have promoted the idea of a 40-mile Backbone Trail crossing the Santa Monica Mountains from Will Rogers State Historic Park to Point Mugu State Park. When completed, the trail will link the three large state parks of Topanga, Malibu, and Point Mugu, enabling Southland residents to spend days and weekends hikng and backpacking across the spine of the Santa Monica Mountains. Trail camps and hostels will eventually be established.

The project has moved along quite slowly for political and financial reasons and each delay means the private land that must be purchased increases in price. A trail corridor ranging from a few hundred yards to a quarter mile wide must be acquired in order to protect the ridges and canyons en route. A major gap in the Backbone Trail is between Topanga and Malibu State Park (where trail camps will be located).

It's the author's hope and the dream of many hikers that feeder trails will one day link the Backbone Trail to population centers in the San Fernando Valley and to the California Coastal Trail in Malibu. CCT will eventually make use of the entire Backbone Trail as the primary route through L.A. County. For now, dayhikers can enjoy parts of the Backbone Trail and join conservationists in the struggle to preserve some much needed open space.

One completed part of the Backbone Trail departs from quiet and imperturbable Topanga Canyon, surrounded by L.A. sprawl, yet retaining its wilderness character. The trail is a good fire road and passes through giant rye grasses and ceanothus on the way to Eagle Rock and Eagle Spring. A longer one-way option takes you along brushy ridges to Will Rogers State Park.

Directions to trailhead: From Topanga Canyon Boulevard, turn east on Entrada Road; that's to the right if you're coming from Pacific Coast Highway. Follow Entrada Road by turning left at every opportunity until you arrive at Topanga State Park. (Be careful if you're coming down Topanga Canyon Blvd. from the San Fernando Valley side of the Santa Monicas. There's an Entrado Drive on the west side of the road that does *not* take you to the park.) The trailhead is at the end of the parking lot.

To Will Rogers Park Trailhead: From Sunset Boulevard in Pacific Palisades, turn north at the park entrance. The road leads up to Rogers' estate. Near Will Rogers' home, a marked trail takes off leisurely uphill on the Inspiration Loop toward Inspiration Point. Rogers Trail intersects 1/10 mile east of Inspiration Point cutoff.

Note: As there is an 1800 foot elevation loss from the top of the fire road in Topanga State Park to Will Rogers Park, it's much easier hiking from Topanga to Will Rogers than vice versa. Purists who want to begin at the official Backbone Trail trailhead in Will Rogers will relish the challenge of a hearty climb.

The Hike: From the Topanga State Park parking lot, follow the distinct trail eastward to a junction, where you'll make a left turn (north) onto Eagle Springs Road. You'll pass through an oak woodland and through chaparral country singed by a 1977 fire. You'll slowly and steadily gain about 800 feet in elevation on the way to Eagle Rock. Enjoy the view of the ocean to the west and the San Gabriel Mountains to the east. When you reach a junction, bear left on the north loop of Eagle Spring Road to Eagle Rock (2104'). A short detour will bring you to the top of the rock, its caves, and an impressive view of Santa Ynez Canyon.

To complete the loop, bear right (south) at the next junction, following the fire road as it winds down to Eagle Spring. The spring includes two wooden water tanks and a pipe from which water drips and nurtures a patch of poison oak. When the Backbone Trail is completed, a trail camp will be built at Eagle Springs. Past the spring, you return to Eagle Spring Road and retrace your steps to the parking lot.

Option: To Will Rogers State Park. Follow the loop trip's directions to the north end of Eagle Rock/Eagle Spring Loop, where you bear right on Fire Road 30. In ½ mile you reach the intersection with Rogers Road. Turn left and follow the road (really a trail) for 3½ miles, where the road ends and meets Rogers Trail. Here a large level area and a solitary shady oak suggest it's time for a lunch break. On clear days relish the spectacular views from every direction: to the left is Rustic Canyon and the crest of the mountains near Mulholland Drive. To the right, Rivas Canyon descends toward the sea.

Don't take the overgrown fire road curving off to the right, but stay on Rogers Trail, which marches up and down several steep hills for about two more miles until it enters Will Rogers Park near Inspiration Point.

Malibu Creek ("The Gorge")

Backbone Trail #2 (LA-5)

Tapia Park-Malibu Creek State Park Loop
12 miles RT; 2000' gain

This hike utilizes another section of the Backbone Trail, which one day will cross the Santa Monica Mountains from Will Rogers State Historic Park to Point Mugu. Hikers follow well-graded fire roads, heading first west along Mesa Peak Motorway, climbing toward Castro Crest, then north and east through Malibu Creek State Park. Fine ocean and island views are offered along the first half of the hike and a chance to explore geologically and ecologically unique Malibu Creek Canyon is yours on the second half.

Directions to trailhead: From Pacific Coast Highway, turn inland on Malibu Canyon Road, proceeding a little over 5 miles to the unsigned road leading to the wastewater treatment plant. This road is just south of the bridge over Las Virgenes Creek. (If you're traveling from the north on Malibu Canyon Road, the trailhead is ¼ mile beyond Tapia Park.) A turnout for parking is located just south of the road to the treatment plant below a slope dotted with cement building foundation slabs. The unsigned trailhead is 200 feet south of the parking area on the west side of Malibu Canyon Road near a park boundary sign.

The Hike: Ascend on the steep dozer trail over brushy slopes. Soon the trail forks. Both forks lead to Mesa Peak Motorway; however, the right fork adds an extra ½ mile to your journey and passes the not-very-inspiring sewage treatment plant. Bear left to the top of a minor ridge. A trail on the spine of this ridge leads 100 yards to Mesa Peak Motorway. Turn left and begin a moderate-to-steep ascent. Handsome oaks shade the road. With the elevation gain come sweeping panoramic views of Point Dume, Santa Monica Bay, and Palos Verdes Peninsula. On clear days, Catalina, San Clemente, Anacapa and Santa Cruz Islands float upon the horizon.

About 2 miles from the trailhead, where Mesa Peak Motorway makes a hairpin turn, an unsigned jeep road veers left toward Mesa Peak (1844'). Our trail continues climbing in a northwesterly direction through an area rich in fossilized shells. Hillside roadcuts betray the Santa Monica Mountains' oceanic heritage. As you hike the spine of the range, a good view to the north is yours: The volcanic rocks of Goat Butte tower above Malibu Creek gorge, the path of Trifuno Canyon can be traced, Malibu Lake rnando Valley over this ridge to the sea. Several thousand acres of Malibu Creek State Park brushland were charred. However, the comeback of a chaparral community as it rises phoenix-like from the ashes is a wonder to behold. Manzanita buds, yucca blossoms, chamise regenerates and the first spring after a fire brings superb wildflower displays.

The road descends briefly to an area of interesting sandstone formations, defaced by "No Trespassing" signs afixed to the rocks by Coralibu Ranch. The road soon intersects paved Corral Canyon Road. Bear right for ⅛ mile. The road becomes dirt, a sign welcomes you back to state park property, and the ascent resumes. Another mile's travel brings you to a road junction signed by painted pipes. Backbone Trail continues west toward the forest of antennas atop Castro Peak (2824') and hopefully, some day, to Point Mugu. Our hike bears right on Bulldog Motorway.

The road descends steeply along the boundary of newly-acquired National Park Service land and winds under electrical transmission lines. The Motorway veers east after 1½ miles and drops into Trifuno Canyon. The road crosses Fern Creek, passes a locked gate, and intersects 20th Century Road. To the left you'll spot where Fern Creek joins Malibu Creek, usually wiping out 20th Century Road in the process during the rainy season.

Turn right on 20th Century Road, soon passing the abandoned set for the M*A*S*H TV series. Stay with the fire road paralleling Malibu Creek. After a short distance, rockhop the creek, and continue on the road. A short road branching off to your right leads to Century Lake, a man-made lake scooped out by members of Crag's Country Club, a group of turn-of-the-century businessmen who had a nearby lodge. Near the lake are hills of porous lava and topsy-turvy sedimentary rock layers telling of the violent geologic upheaval that formed Malibu Canyon.

The road descends down a hill and you may proceed either on the High Road on the north side of the creek or cross the bridge over the creek and hike along the south side. The park Visitor Center occupies the big white house near the bridge crossing. The dirt road becomes asphalt and you soon reach the park entrance kiosk and parking area. Continue out of the park to Malibu Canyon Road. Turn right, following the road for 1¼ miles back to the trailhead.

Zuma-Dume Trail (LA-6)

Zuma Beach to Pt. Dume; 2 miles RT
Zuma Beach to Paradise Cove; 4 miles RT

Zuma Beach is L.A. County's largest sand beach and one of the finest white sand strands in California. Zuma lies on the open crest beyond Santa Monica Bay and thus receives heavy breakers crashing in from the north. From sunrise to sunset, board and body surfers try to catch a big one. Every month the color of the ocean and the Santa Monica Mountains seem to take on different shades of green depending on season and sunlight, providing the Zuma Beach hiker with yet another attraction.

Zuma's wealth of sand is inherited from material dumped into the Pacific by Trancas and Zuma Creeks. Pt. Dume helps also by acting like a gigantic groin, capturing sand along its up-current side. Offshore, the slope of the sea bottom may influence wave refraction and thus dump even more sand on the beach.

This hike travels along the western part of Zuma Beach, climbs over the geologically fascinating Pt. Dume Headlands for sweeping views of the coast, and descends to Paradise Cove, site of a private beach and fishing pier.

Directions to trailhead: Zuma Beach County Park is located at the 30000 block of Pacific Coast Highway, Malibu. There is a parking-admission fee. Although the east beach is usually crowded in the summer months, the west beach is not. Find a parking space near the west end and begin your hike. Mass transit: RTD #175.

The Hike. Walk down-coast along sandy Zuma Beach. This stretch of Zuma is known as Westward Beach. The water here is clearer and colder than Santa Monica Bay to the east of Pt. Dume. A little inland from the beach is Point Dume Whale Watch, reached by a stairway from Westward Beach Road. In winter the possibility of seeing a migrating gray whale swimming south toward Baja is good. The migration route brings them quite close to shore.

On the western side of Point Dume is Pirate's Cove, two hundred yards of beach tucked away between two rocky outcroppings. It's the scene of much dispute between nude beach advocates, residents, and the county sheriff. The state is currently acquiring Pirates Cove and other acreage along the Pt. Dume Headlands as an ecological reserve.

Option to Paradise Cove: Take the sandy goat trail, which leads up the sandstone cliffs to the top of Pt. Dume. Wander atop the point and observe the white sea cliffs of Santa Monica arcing inland. As you stand atop the rocky triangle projecting into the sea, observe the dense blac kZuma volcanics and the much softer white sedimentary beds of the sea cliffs extending both east and west. The volcanics have resisted the crashing sea far better than the sedimentary rock and have protected the land behind from further erosion, thus forming the triangle shape of the point.

Keep toward the edge of the bluffs, following more goat trails until you spot one that descends the east side to the beach. A bit of beach walking brings you to Paradise Cove, sheltered from the north and northwest by Point Dume and the souht by extensive kelp beds. You can rent a boat and fishing tackle and go fishing around the kelp beds. Diving is good here, too. Paradise Cove, sometimes called Dume Cove, is a romantic spot and the scene of much TV and motion picture filming.

Return the same way.

Leo Carrillo Trail (LA-7)

Leo Carrillo State Beach to County Line; 3 miles RT
Leo Carrillo State Beach to Sequit Ridge; 6 miles RT

The state beach is named after Angeline Leo Carrillo, famous for his TV role as Pancho, the Cisco Kid's sidekick. Carrillo was also quite active in recreation and civic matters.

Leo Carrillo Beach is stabilized to some extent by minor rocky breaks in the sandy shoreline and by extensive kelp beds offshore. Seals come ashore frequently. (Don't disturb.)

This trail follows one of L.A. County's more interesting and natural beaches. At Sequit Point you'll find good surfing, swimming, skin diving and a cluster of caves and coves. An optional trail allows you to head inland, climing Arroyo Sequit Ridge for panaromic views of the Channel Islands.

Directions to trailhead: Leo Carrillo State Beach is located just downcoast from Mulholland Highway on Pacific Coast Highway near the L.A.-Ventura County Line. Park along PCH (free) or at the state beach (fee).

The Hike: Head up-coast toward Sequit Point. The point bisects the beach, forming a bay to the south. Surfers tackle the well-shaped south swell, battling the submerged rocks and kelp beds.

As you near the point, you'll pass a path which leads under the highway and connects the beach with the sycamore-shaded campground. The stae got a bit carried away with pouring asphalt at Leo Carrillo, but it's still a nice park. Scramble around the rocks of Sequit Point, to several rock formations, caves, coves, a rock arch, and some nice tidepools.

North of the point Leo Carillo offers good swimming with a sandy bottom. The unspoiled coast here is contrasted with development in the county line area. When the beach narrows nearly to the point of disappearing and the condos multiply, return the same way.

Option: Yellow Hill Fire Road to Sequit Ridge. The fire road begins near the ranger residences on the west side of Mulholland Highway. For one third of a mile, you'll parallel the coast, then turn inland, climbing all the while. You can see several Channel Islands: Anacapa, Santa Cruz, Santa Rosa. On an especially clear day, you can sometimes see miniscule Santa Barbara Island due south.

The fire road passes through sage and chaparral to a 1600-foot peak. This is the northwest boundary ofthe park where you turn around and return the way you came.

Triathlon Trail (LA-8)

Santa Monica Pier to Torrance County Beach; 19 miles one way

The traithlon, an endurance competition requiring participants to run, bicycle and swim long distances, has become popular among the super-jock set in recent years. This beach triathlon lets you exhaust every muscle in your body and enjoy a beautiful stretch of Southland beach. Our beach triathlon, however, can be tailored to your own physical fitness.

The beach between Torrance and Santa Monica is wide and sandy, interrupted only by King Harbor Marina and Marina del Rey. Along the beach is the South Bay Bicycle Path, which Undulates south from Santa Monica to Torrance Beach. Ride as much of the bike trail down-coast as you wish, return up-coast on foot. Then, to complete your personal triathlon, take a long refreshing swim.

Directions to trailhead: The Santa Monica Pier is located at Colorado and Ocean Avenue. There's parking in the beach lots nearby. Mass Transit: RTD #'s 75, 85, & 175; Santa Monica MLB #'s 1, 7, & 10. Torrance County Beach at the other end of the Triathlon Trail is accessible through the parking lot at the north end of Redondo Beach. Turn oceanward from Pacific Coast Highway in Redondo Beach. Mass Transit: RTD #'s 846, 867, 871.

The Hike: (See L.A. County CCT description for detailed account of this trail.) Most of Los Angeles' urban waterfront offers clean sand, moderate surf, and safe swimming. Volleyball, surfing and countless snack bars, sidewalk cafes and raft-renting establishments can be found here. Frequent bus service runs along these Southland beachs. The beach towns line the sane in Mediterranean style, the boardwalks bustle with friendly folk.

The eight miles of South Bay Beaches extending south of Marina del Rey to the north end of Palos Verdes—Playa del Rey, Manhattan, Hermosa, Redondo, Torrance—are especially nice.

Ventura County

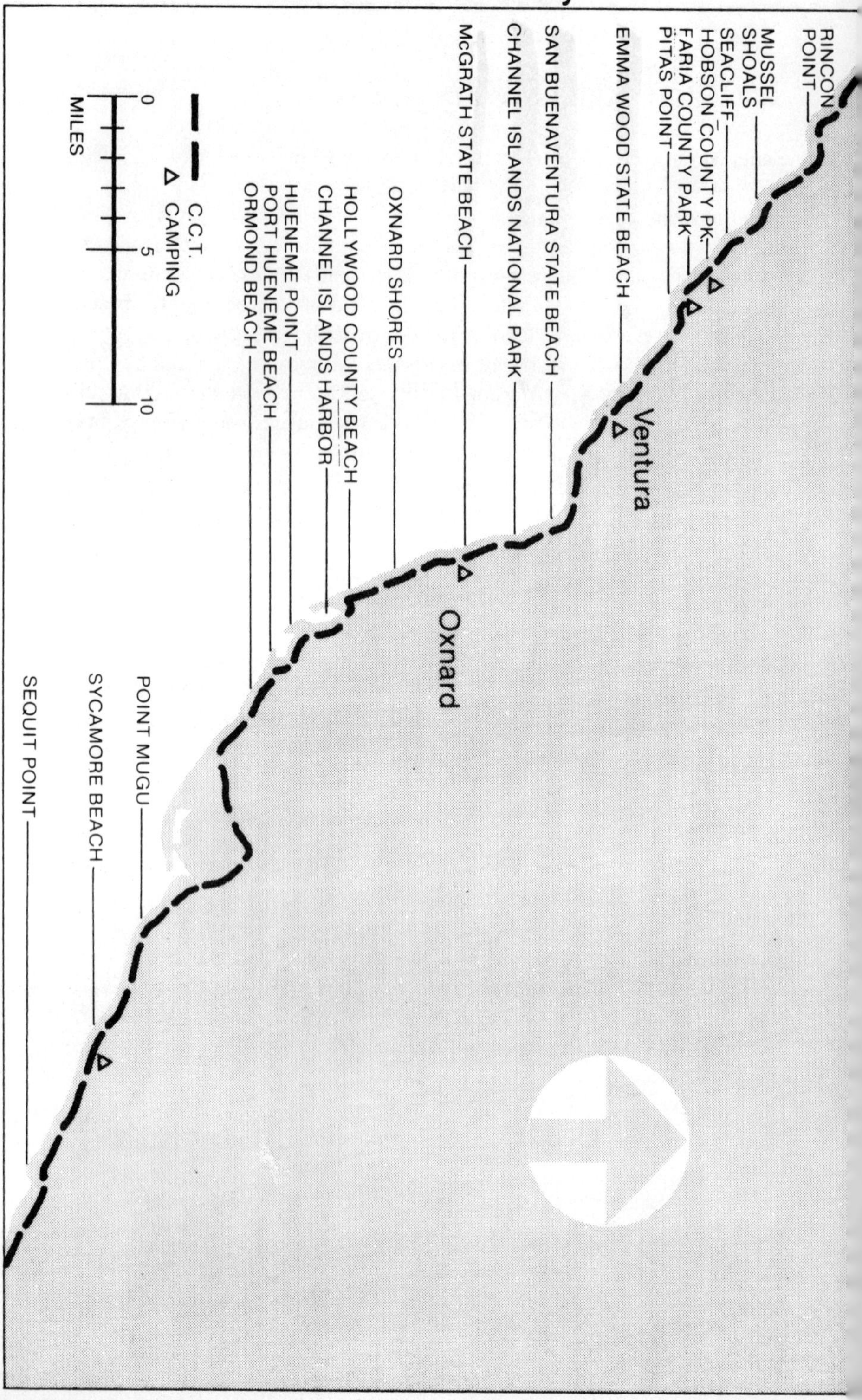

- RINCON POINT
- MUSSEL SHOALS
- HOBSON COUNTY PK.
- SEACLIFF
- FARIA COUNTY PARK
- PITAS POINT
- EMMA WOOD STATE BEACH
- Ventura
- McGRATH STATE BEACH
- CHANNEL ISLANDS NATIONAL PARK
- SAN BUENAVENTURA STATE BEACH
- OXNARD SHORES
- Oxnard
- HOLLYWOOD COUNTY BEACH
- CHANNEL ISLANDS HARBOR
- HUENEME POINT
- PORT HUENEME BEACH
- ORMOND BEACH
- POINT MUGU
- SYCAMORE BEACH
- SEQUIT POINT

MILES 0 — 5 — 10

– – – C.C.T.
△ CAMPING

Ventura County

Ventura County's 43 miles of shoreline offer several fine sandy beaches; the best ones are the state parks, Point Mugu, McGrath, San Buenaventura and Emma Wood. Rising above the beach are the Santa Monica Mountains in the south county and the Santa Ynez Mountains in the north county. Between these two ranges is a broad fertile agricultural delta, the Oxnard Plain. Strawberries and citrus grow here, as well as the spreading suburbs of Oxnard and Ventura.

CCT hikers will find it rocky going in the southern and northern parts of the county. Ventura County's central coast is sandy beach deposited by the former delta area of the Santa Clara River.

Point Mugu Lagoon, one of Southern California's most pristine wetlands is owned by the Navy, which permits no trespassing. The lagoon is entirely within the borders of the Pacific Missile Test Center at Point Mugu.

Ventura Harbor is the location of Channel Islands National Park Headquarters: 1699 Anchors Way Drive, Ventura, CA 93003, (805) 644-8157. Visitors come to San Miguel and Anacapa, Santa Barbara and Santa Cruz Islands to view the sea elephants, enjoy the giant coreopsis in bloom, and watch the many sea birds and whales in migration.

Ventura beaches are rougher and more exposed to the elements than others in Southern California because the coastline faces more nearly west. However, in summer and on warm winter days they can be as pleasant as any.

> **CCT at a glance**
>
> **Terrain:** Rocky in northern and southern parts of the county; sandy in the middle. Rincons (rounded points) in north county better for surfers than hikers.
>
> **Obstructions:** Navy's Pacific Missile Center at Pt. Mugu, naval Construction Batallion Center at Point Hueneme, Channel Islands Harbor, Ventura Harbor.
>
> **Transportation:** South Coast Area Transit (SCAT) 301 E. 3rd St., P.O. Box 1146, Oxnard 93032
>
> **Campgrounds-Accommodations:**
> Pt. Mugu State Park
> McGrath State Park
> Emma Wood State Beach
> Faria County Park
> Hobson County Park

CCT: Ventura County

CCT follows Leo Carrillo Beach north into Ventura County. The beach grows increasingly rocky, forcing hikers closer and closer to the highway. At the county line, you'll find some residential development, but further on, County Line Beach is undeveloped and a popular surfing spot.

As you cross into Ventura County, the view ahead is of a rugged rocky coast with mountain slopes descending precipitously into the ocean. The marine terraces we've seen to the south disappear beneath the sea here, owing to the westward tilting of the land. Obviously it was hard work constructing the highway here; hiking is hard work too.

After three miles of rock-hopping, CCT enters Point Mugu State Park on Sycamore Cove Beach. A nice grassy picnic area is next to the beach. Opposite the beach, on the other side of Highway 1, is Sycamore Canyon Campground with drive-in campsites and a special biker-hiker campsite. The trailhead at the campground points hikers up the beautiful sycamore-lined canyon. The trailhead marks the terminus (or beginning) of the Backbone Trail, the long trail running along the spine of the Santa Monica Mountains from Will Rogers State Park to Point Mugu State Park. It's a

4-mile hike up Sycamore to La Jolla Valley Walk-in Campground. An enjoyable loop can be made by hiking up Sycamore Canyon, spending the night in La Jolla Valley and returning to the coast on the trail through La Jolla Canyon.

Soon you'll notice a large sand dune on the landward side of the highway. Dune riders slide down the sand on old skis and surfboards, pieces of cardboard and inflatable surf-riders. The dune is formed by prevailing westward winds, which scoop up beach sand, carry it across the highway, and hurl it against the mountains. Wind is one of the few processes in nature with the ability to transport material uphill.

A mile up-coast from Sycamore Cove Beach, CCT reaches La Jolla Beach Campground, 102 trailer and tent campsites on the beach next to Highway 1. It has all the charm of a landing strip. Tent sites squat at the north end of the beach. Hikers would be better advised to camp in the Sycamore Canyon hiker-biker campsite or walk an easy two miles on signed trail up La Jolla Canyon to the La Jolla Valley walk-in campground.

Ahead looms a huge pyramid-shaped rock knob, Pt. Mugu. Around the tip of the point, where waves smash against the great rock, you can see the remains of the coast highway, which once rounded the point. A deep slot now allows motorists through the middle of the rock. The point is rich in fossil shell fragments and spray-can graffiti.

On a clear day, you can see Santa Cruz and Anacapa Islands. When you observe the Channel Islands from Point Mugu, the theory the islands' originated as a westward extension of the Santa Monica Mountains seems credible, though geologists argue this point.

Mugu is believed to have been derived from a Chumash word, *muwu,* meaning beach and also refers to a village near here. Cabrillo mentioned this village during his 1542 exploration and speculation has it that Pt. Mugu may be the oldest recorded California place name still in existence.

Beyond the point, small sandy Point Mugu Beach extends to the Point Mugu Pacific Missile Range. The Navy has a small arms firing range at the south end of the Range; the beach is closed to the public. CCT hikers must follow the highway from here around Mugu Lagoon, the only undisturbed estuarine area left in Southern California. The lagoon was formed when waves deposited the sand bar that now separates the brackish wetland from the open sea. The lagoon is a pristine aquarium of marine life. Public access is prohibited.

CCT hikers should be warned that the Missile Range and navy Construction Battalion Center farther north add up to over 12 miles of unpleasant road-walking. The next few paragraphs describe the Naval Dodge; however, it's the author's recommendation that hikers dodge this entire section of coast and resume hiking the California Coast Trail at

Hollywood County Beach at the end of Channel Islands Boulevard. From here, sixty miles of beautiful, relatively uninterrupted, shoreline stretches to Gaviota.

One mile past the Vista Point at Mugu Lagoon, hikers cross Calleguas Creek at the highway bridge and merge onto wide, flat Oxnard Plain. Fruits and vegetables are cultivated on this fertile plain. Bear left on Hueneme Road, then left on Arnold Road which returns you to the beach.

At Ormond Beach is a wetland and sand dune area popular with ORV enthusiasts on three-wheeled motorcycles and dune buggies. Dominating the industrial shoreline are the tall stacks of Edison Company's Ormond Beach generating station. Power lines extend across the Oxnard Plain, borne by giant towers.

Ormond Beach gives way to Port Hueneme Beach, a large sandy beach with a fishing pier. A wharf was completed near here in 1871, the first built between San Pedro and Santa Cruz. For years, Hueneme was one of the largest grain-shipping ports on the Pacific Coast. In the late 1930s, a port was dredged into the original Hueneme Lagoon. During World War II, the Navy purchased the entire harbor. You may walk out along the wave-battered jetty to the harbor entrance (cover your ears when the fog horn sounds), but there's no passage around the harbor due to the U.S. naval Construction Batallion Center, "Home of the Seabees."

A short distance beyond Port Hueneme Pier is Ventura Road; take it two miles north to Channel Islands Boulevard, go left (west) another two miles to the beach.

CCT becomes interesting and relaxing again at Hollywood Beach, named for a residential community, Hollywood-By-the-Sea, which sprang up after Rudolph Valentino's *The Sheik* was filmed on the dunes here. A mile's hike north brings you past the private community of Oxnard Shores, famed for getting clobbered by heavy surf at high tide. The beach is flat and at one time was eroding at the phenomenal rate of 10 feet a year. Homes are built right on the shoreline and many have been heavily damaged. New homes are built on pilings, so the ocean crashes under rather than through them. North of Oxnard Shores is the county's second coastal Edison plant, this one a relic from the '50s. Mandalay County Park here features some nice dunes and an undeveloped beach.

A short walk over dunes brings you to McGrath Lake. Many species of birds, including white-tailed hawks and herons congregate around this small lake at the south end of McGrath State Beach. The lake and the beach extending north to the Santa Clara River are protected by McGrath State Beach. Surf fishermen cast for corbina, bass and perch along the state beach. McGrath campsites are located on the edge of a salt marsh near the mouth of the Santa Clara River. Sites are somewhat protected from westerly winds by low sand dunes.

CCT follows a narrow trail (difficult to spot from the beach) along the Santa Clara River, heading inland. Muskrats, rabbits, turtles, gopher snakes, and many species of native and migratory waterfowl make their home in the protected marsh known as the Santa Clara Estuary Natural Preserve. The river trail takes you through lush vegetation, past a wastewater treatment plant (you'll smell it before you see it) to Harbor Boulevard.

Bear north on Harbor across the Santa Clara River Bridge, continuing a half mile to Ventura Harbor. The harbor was washed out by the 1969 Santa Clara River flood, but is back bigger than ever. CCT hikers improvise a route around the boat basins. If you wish, follow the signs to Channel Islands National Park Headquarters, 1699 Anchors Way Drive, Ventura 93003, (805) 644-8157. Rangers will tell you about boat trips to the islands.

CCT follows the boat basins on residential streets around the marina.

North of the marina, CCT arcs around Pierpont Bay to San Buenaventura State Beach, long developed as a residential-recreational area. The state beach extends for a mile along the city's waterfront to beyond the 1700-foot fishing pier. The wide sandy beach is protected by breakwaters and offers good swimming.

> "It does not seem to be generally known, nor much appreciated by home people, that Ventura as a watering place has no superior, if equal on the Pacific Coast. Aside from possessing a climate the most delightful and healthful, it has a beautiful beach stretching for miles along the waterfront. There is not time when one cannot enjoy a drive on it a distance of from 3 to 6 miles and when the tide is out, 10 or 12 miles on the edge of the foaming and curling surfline over a roadbed as hard and smooth as planed oaken floor."
> —*Ventura Democrat* 1885

The waterfront property along the state beach and the county fairgrounds to the north is heavily developed with huge high rise apartments and hotels. However, the development soon vanishes as CCT approaches Surfer's Point Park. The beach here is narrow and rock-rimmed. It's a fine place to learn to surf. Up-coast, the beach grows increasingly cobbled and hikers rock-hop across Ventura County Fairgrounds Beach or walk on the low bluffs of the fairgrounds.

Ford the Ventura River, usually deep only during flood season. The wetlands here was known as Hobo Jungle. In bad weather hobos slept under the railroad bridge that crosses the river. A century ago the river was a wild place; speckled trout were pulled from its waters, rabbits and quail shot along its banks.

The coast between the Ventura River and Rincon Point, 13 miles ahead, is called the Rincon although it could be termed the Rincons, plural, because there are three indentations in the coast, separated by Pitas Point and Punta Gorda. Geologically-minded hikers will find exposures of Modelo, Pico, Santa Barbara and San Pedro formations aong the Rincons.

Across the river, CCT enters Seaside Wilderness Park, an undeveloped melange of sand dunes, Monterey pine and palm trees. It's a delight for birdwatchers. Just ahead is sandy and cobbled Emma Wood State Beach, facing a reef that made it a favorite winter big-wave spot in surfing's early days. Surfers still catch some big ones here. The campsites, protected by a wind barrier, lie between the Southern Pacific railroad tracks and a wave-splashed seawall. A group hiker-biker campground is located at the south end of the park.

Two miles up-coast from Emma Wood, you pass the tiny community of Solimar and begin arcing west to Pitas Point. Extensive kelp beds grow a few hundred yards offshore. In summer you may see beds of white Marguerite blooming on the terraced surfaces off to your right. The point's name began as Pitos, Spanish for "whistles" and was named by the Portola expedition in 1769. Diarist Father Crespi complained: "During the night they [the Indians] kept us awake, playing all night on some doleful pipes or whistles." Confused mapmakers later changed the spelling, thinking the name came from century plant, which is *pita* in Spanish. The park at Pitas Point was donated in 1915 by Manuel Faria, a native of the Azore Islands. Faria County Park is a small day use and camping spot.

Two more miles bring you to Hobson County Park, named for William Dewey Hobson, called the "Father of Ventura County" because he was instrumental in securing the passage of legislation that created Ventura County in 1873. The park has oceanfront campsites. Next to the park is the tiny community of Sea Cliff. To protect their homes, residents have hauled in huge boulders and contructed sea walls.

Sea Cliff and Hobson County Park lost their beach when the state began constructing the new Highway 101 route along the Rincon. The highway project dumped 2.8 million cubic yards of dirt into the ocean. This new man-made "coastal bluff" stopped the littoral flow of sand down the coast and wiped out the waterfront here.

Again CCT arcs west with the second mini-rincon and passes the small community of Mussel Shoals, where this mollusk abounds, and sea shells are encrusted on a projection of resistant rock. CCT rounds Punta Gorda. Gorda means "big" or "fat" and well describes the massive promontory here.

A pier extends from the point to Richfield Island, an artificial island covered with palm trees and oil wells.

CCT rounds the point and crosses small, sandy La Conchita Beach. On the inland side of the highway is La Conchita Del Mar, a sea colony settled in the 1920s. The euphonious name means "small sea shells."

CCT follows the third and final rincon on its three-mile-long curve to Rincon Point.

As CCT nears Rincon Point, the ocean closely approaches Highway 101, especially at high tide when waves crash against the protecting rock embankment. For many years this stretch of coast was a barrier for travelers because one could pass only at low tide on the wet sands of the beach; at high tide, the waves dashed against the white cliffs of the mountains. Don Gaspar de Portóla on his way north in search of the Bay of Monterey, Father Junipero Serra as he traveled from mission to mission on foot, and John C. Fremont and his buckskin mountain men during the conquest of California all had to wait until the tide receded before they could pass.

In the early days of automobiling, a wooden plank causeway was built over the water. Columns of water occasionally shot through holes in the boards, startling horseless carriage drivers. Children shrieked in delight, though it proved distracting to dad.

Rincon Point is made by the fan delta which built a sweeping spur into the ocean at the north end of Rincon Creek. The creek isn't long, but it descends precipitously from the hills, and at flood stage carries boulders to the ocean. The long shore current can't move the large rocks, so they accumulate to make the rocky spur of the point.

East of the point, surfers catch the swells refracted around the point and ride them near-parallel to shore. Some of Southern California's best waves break here in a foamy maelstrom, a true challenge for skilled kayakers and surfers.

As CCT rounds Rincon Point, you enter Santa Barbara County at Rincon Beach County Park.

La Jolla Valley Trail (V-1)

La Jolla Creek to La Jolla Valley Trail Camp, 4 miles RT; 700′ gain
Return via Mugu Peak Trail, 8 miles RT; 1000′ gain

Ringed by ridges, the native grassland of La Jolla Valley welcomes the hiker with its drifts of oaks and peaceful pond. This pastoral upland is unique: it has resisted the invasion of non-native vegetation. It's rare to find native grassland in Southern California because the Spanish introduced oats and a host of other foreign grasses for pasture. In most cases, the imported grasses squeezed out the natives; but not in La Jolla Valley.

This trail tours the beautiful grasslands of the valley and the longer option offers superb views of the Pacific and Channel Islands.

Directions to trailhead: Drive about 30 miles north on Highway 1 from Santa Monica (21 miles past Malibu Canyon Road if you're coming from the San Fernando Valley). The hard-to-spot turnoff is 1½ miles north of Big Sycamore Canyon, which is also part of Point Mugu State Park. From the turnoff, bear right on the narrow road and continue ¼ mile to the parking area. The signed trailhead, near an interpretive display, is at a fire road going into the canyon.

The Hike: The fire road heads north up the canyon through rolling chaparral. In ¾ mile it passes through an area of rock excavation and crosses a branch of La Jolla Creek near some seasonal falls. At the creek crossing, the fire road ends and the trail begins along the steep right side of the canyon. In ½ mile, the trail joins a crumbly dirt road and continues above the long, narrow oak forest that straddles the creek. On the hotter slopes above the oaks reside the sumacs, blackened by fire but unbowed. The trees' dead branches have dropped off, making them appear armless, but a closer look reveals that the empty sockets are sprouting new limbs; a tiny miracle and an example of nature's regenerative powers.

At the first trail junction, bear right on the La Jolla Valley Loop Trail. You'll emerge to the left after completing your tour of La Jolla Valley and circling Mugu Peak. Following the right fork, you'll soon arrive at another junction. Leave the main trail and follow the east fork, you will descend the short distance to a lovely cattail pond. The pond is a nesting place for a variety of birds including the redwing blackbird. Above the pond is a picnic area with tables, running water, and a good view of the ducks.

Return to the main trail. Continue and you'll skirt the east end of the valley with an overview of the waving grasses, and intersect a "T" junction. To the right, ½ mile to a signed junction and a longer connecting trail back to the La Jolla Canyon Loop Trail. Either route is an easy 2 mile return.

Option: Return via Mugu Peak Trail. Continue past La Jolla Valley Walk-in Camp and hike through the billowing grassland. Keep an eye out for speedy roadrunners, who dash through the grass. The trail heads toward the "spy in the sky," the round missle-tracking disc atop Laguna Peak to the west. Proceed straight ahead to an unmarked trail junction. Here a sign indicates you've reached the limits of Point Mugu preservation area. The trail crosses a number of branches of La Jolla Creek, following this watercourse and its oak copses as it cuts through the rolling grassland. You'll want to dawdle near here, perhaps do a little cross country thrashing on one of the deer trails or go off stalking wild chocolate lilies hidden in the grass.

The trail continues south toward the Pacific and soon climbs a low saddle where there are splendid views of the ocean and the Channel Islands. The cool breeze is welcome relief from the hot prairie. The trail contours around the prickly pear and yucca covered slopes of Mount Mugu (1266′). After rounding the mountain, you recross La Jolla Valley Creek, bear to the right, and soon intersect the La Jolla Canyon Loop Trail. Bearing right here, you return to the trailhead.

McGrath Beach Trail (V-2)

McGrath State Beach to McGrath Lake; 4 miles RT
McGrath State Beach to Oxnard Shores; 8 miles RT
McGrath State Beach to Channel Islands Harbor; 12 miles RT

McGrath State Beach and McGrath Lake were named for the McGrath family which had extensive land holdings in the area dating from 1876. The park is situated on the edge of a salt marsh, just south of the Santa Clara River. North of the entrance kiosk is a nature trail.

This hike takes you to the Santa Clara River Estuary, McGrath Lake, and along miles of sandy beach to Channel Islands Harbor.

Directions to trailhead: You may park along Harbor Boulevard just north of the Santa Clara River at McGrath State Beach, located off Harbor Boulevard 1.2 miles south of Spinnaker Dr.

The Hike: If you begin this hike at the Santa Clara River bridge, head oceanward on the sporadic trail through the nature preserve. Along the river bank is a mass of lush vegetation, populated by squirrels, turtles, gopher snakes, rabbits, and many native and migratory waterfowl.

When your reach the dunes piled high near the shore, head down-coast along McGrath State Beach. The beach is clean and natural and backed by low dunes. In two miles you'll spot McGrath Lake, tucked away behind some dunes. Owls, herons and a host of birds visit the lake.

Options to Oxnard Shores: More sandy beach and dunes follow. You pass a power plant and arrive at Oxnard Shores, a development built too close to land's end; famous for getting washed away.

Option to Channel Islands Harbor. A mile of beach walking brings you to historic Hollywood Beach. *The Sheik* starring Rudolph Valentino was filmed on the desert-like sands here. Real estate promoters attempted to capitalize on Oxnard Beach's instant fame and re-named it Hollywood Beach. They laid out subdivisions called Hollywood-by-the Sea and the Silver Strand, suggesting to their customers that the area was really a movie colony and might become a future Hollywood.

This hike ends another mile down-coast at the entrance to Channel Islands Harbor. Return the same way.

Emma Wood Trail (V-3)

Emma Wood State Beach to Seaside Wilderness Park; 3 miles RT

Emma Wood acquired large landholdings on the coast and coastal slope of north Ventura County, land that was originally part of Rancho San Miguelto. She died in 1944 and in the late '50s, her husband and heir gave the state the beach west of the railroad overpass, a stretch of coast long popular with the public.

This hike takes you along Emma Wood's mixed rock and sand beach to Seaside Wilderness Park, a small undeveloped spot near the Ventura River, where pine and palm trees rise above low sand dunes.

Directions to trailhead: Begin this beach hike at Emma Wood State Beach, located at the south end of old Pacific Coast Highway, north of West Main Street in Ventura. Take the Emma Wood exit off Highway 101. There's a fee for day use. You can also park just outside the park and walk under the highway to the group campground at the south end of the park and begin your hike from there.

The Hike: Cross the railroad tracks to the beach and head down-coast on sandy beach. The swimming is okay here, but beware of submerged rocks. In the early days of big board surfing, this was a popular spot, due to an offshore reef which increased wave size. The beach becomes intermittently rocky. Looking inland you see stream alluvium and deltaic deposits of the Ventura River. these alluviums are quite thick and as you would expect, the soil on the Ventura Plain is excellent for agriculture.

A short distance up-coast from the Ventura River is Seaside Wilderness Park, a melange of Monterey pine and palm protruding from low dunes. The birdwatching is superb here.

In the nineteenth century, the Ventura River was a wild mountain waterway offering good trout fishing. In the twentieth century, hobos made camps in the wetlands here. In bad weather residents of Hobo Jungle slept under the railroad bridge that crosses the Ventura River. Today, anglers cast for corbina, bass and perch at the river mouth. There's a small lagoon to explore, populated with a variety of shorebirds.

Return the same way.

Anacapa's Arch Rock

Anacapa Island Loop (V-4)

Season: All Year
Topo: Anacapa Island

2 miles RT from Visitors Center of the Channel Islands

Anacapa Island, closest to the mainland, was called Las Mesitas (Little Tables) by Portola in 1769. Later the Chumash Indians name for the island, Eneeapah (believed to mean "deception") came into popular use and the name evolved into Anacapa. Ventura sheepmen once owned the island. There's no water on the island, so it's hard to imagine how the sheep survived. The popular belief is that night fog was so dense that the sheep's coats became soaked at night, each sheep becoming a wooly sponge by morning. So what could be more natural than one sheep drinking from the fleece of another? So the story goes...

Anacapa, just 12 miles southwest of Port Hueneme, is the most accessible Channel Island. It offers the day hiker a sampling of charms of the larger islands to the west. Anacapa also provides a refuge for schools of fish. Below the tall wind-and-wave-cut cliffs, sea lions bark at the crashing breakers. Gulls, owls, herons, and pelicans call the cliffs home.

Anacapa is really three islands chained together with reefs that rise above the surface during low tide. West Anacapa is the largest segment, featuring great caves where Indians are said to have collected water dripping from the ceiling. The middle isle hosts a wind-battered Eucalyptus grove. The east isle, where the National Park Service has its visitors center, is the light of the Channel Islands; a Coast Guard lighthouse and foghorn warns ships of the dangerous channel. Anacapa's 60,000 candlepower light has been in service since 1932. The guardian light is visible for twenty-four miles in all directions and the foghorn never ceases its call.

This day hike tours east Anacapa Island. The island is barely a mile long and a quarter mile wide, so even though you tour the whole island, it's a short hike. The route follows the park service's figure-eight shaped nature trail, which explain some of the human history of the isle and gives you views of Cathedral Cove, a western gull rookery, and miles of blue Pacific.

Directions to trailhead: For the most up-to-date information on boat departures, contact Channel Islands National Park, 1699 Anchors Way Drive, Ventura 93003 (805) 644-8157. Most commercial tour operators leave from the Ventura and Oxnard marinas. A number of "whale watching" tours are offered during winter when the gray whales migrate. Some of these tour boats land on Anacapa Island and some don't, so charter your boat with care. One company, Island Packers, runs year-round to Anacapa. Their boats leave every Saturday and Sunday. Weekday trips are scheduled during whale season and in summer. To contact them: Island Packers Company, P.O. Box 933, Ventura 93001 (805) 642-1393, 642-3370. Another organization that runs trips to the island is Cabrillo Marine Museum, 3720 Steven White Drive, San Pedro, CA 90731, (213) 831-0062.

The Hike: It's a romantic approach to East Anacapa, as you sail past Arch Rock. As you come closer, however, the island looks God-forsaken; not a tree in sight. But as you near the mooring at the east end of the isle, the honeycomb of caves and coves is intriguing and you feel more interested. A skiff brings you to the foot of an iron stairway. You climb 150 stairs, ascending steep igneous rocks to the cliff tops.

What you find on top depends on the time of year. January through March you may enjoy the sight of 30-tons gray whales passing south on their way to calving and mating waters off Baja California. In summer, the isle's vegetation is a dormant brown. Even the giant coreopsis, the island's featured attraction, is a mass of withered leaves and flowers. However, in early spring, the coreopsis is something to behold. It is called the tree sunflower, an awkward, thick-trunked perennial, sometimes reaching ten feet in height. It sprouts large yellow blossoms. When you approach the island by boat and look up at the cliffs, it's as if someone threw a yellow tablecloth atop them.

The nature trail leaves from the visitors' center where you can learn about island life, past and present. A helpful pamphlet is available which describes the trail's features. Remember to stay on the trail. The islands parched ground cover is easily damaged.

Along the trail, a campground and several inspiring cliff-edge nooks invite you to picnic. The trail loops in a figure-eight through the coreopsis and returns to the visitors' center.

San Miguel Island

San Miguel Island Trail (V-5)

Cuyler Harbor to Lester Ranch 3 miles RT; 715 foot gain

San Miguel is the westernmost of the Channel Islands. Eight miles long, four miles wide, it rises as a plateau, 400 to 500 feet above the sea. Wind-driven sands cover many of the hills and beaches which were severely overgrazed by sheep during the island's ranching days. Once owned by the U.S. Navy which used it as a bombing site and missile tracking station the island is now managed by the National Park Service.

Three species of cormorants, storm petrels, Cassin's auklets, and the pigeon guillemot nest on the island. San Miguel is home to six pinniped species: California sea lion, northern elephant seal, steller sea lion, harbor seal, northern fur seal and Guadalupe fur seal. The island may host the largest elephant seal population on earth. As many as 15,000 seals and sea lions can be seen basking on the rocks during mating season.

A trail runs most of the way from Cuyler Harbor to the west end of the island at Point Bennett, where the pinniped population is centered. The trail passes two round peaks, San Miguel Peak and Green Mountain and drops in and out of steep canyons to view the lunar landscape of the caliche forest. Please check in with the resident ranger and stay on establshed trails because the island's vegetation is fragile.

Directions to trailhead: Plan on a long day or overnight trip to San Miguel. It's at least a five-hour boat trip from Ventura. There are no regularly scheduled trips to the island. The Cabrillo Art Museum in San Pedro and the Santa Barbara Natural History Museum sometimes sponsor trips. Contact Island Packers Company, P.O. Box 993, Ventura (805) 642-1396 or Channel Islands National Park Headquarters (at the Ventura Harbor), 169 Anchors Way Drive, Ventura 93003, (805) 644-8157.

The Hike: Follow the beach at Cuyler Harbor to the east. The harbor was named after its original government surveyor in the 1850s. The beach around the anchorage was formed by a bight of volcanic cliffs that extend to bold and precipitous Harris Point, the most prominent landmark on the San Miguel coast.

At the east end of the beach, about ¾ of a mile from anchoring waters, a small footpath winds its way up the bluffs. It's a relatively steep trail following along the edge of a stream-cut canyon. At the top of the canyon, the trail veers east and forks. The left fork takes you a short distance to the Cabrillo Monument.

Juan Rodriguez Cabrillo, Portuguese explorer, visited and wrote about San Miguel in October 1542. While on the island he fell and broke either an arm or a leg (historians are unsure about this). As a result of this injury he contracted gangrene and died on the island in January 1543 and it's believed (historians disagree about this too) was buried here. In honor of Cabrillo, a monument was erected in 1937.

The right fork continues to the remains of a ranch house. Of the various ranchers and ranch managers to live on the island, the most well-known were the Lesters. They spent 12 years on the island and their adventures were occasionaly chronicled by the local press. When the Navy evicted the Lesters from the island in 1942, Mr. Lester went to a hill overlooking Harris Point, in his view the prettiest part of the island, and shot himself. Within a month his family moved back to the mainland. Not much is left of the ranch now. The buildings burned down in the 60s and only a rubble of brick and scattered household items remain.

For a longer 8-mile roundtrip the hiker can continue on the trail past the ranch to the rop of San Miguel Peak (916′) down and then up again to the top of Green Mountain (850′). Ask rangers to tell you about the caliche forest composed of calicified sheaths of plants that died thousands of years ago. Calcium carbonate has reacted with ancient plants' organic acid, creating a ghostly forest.

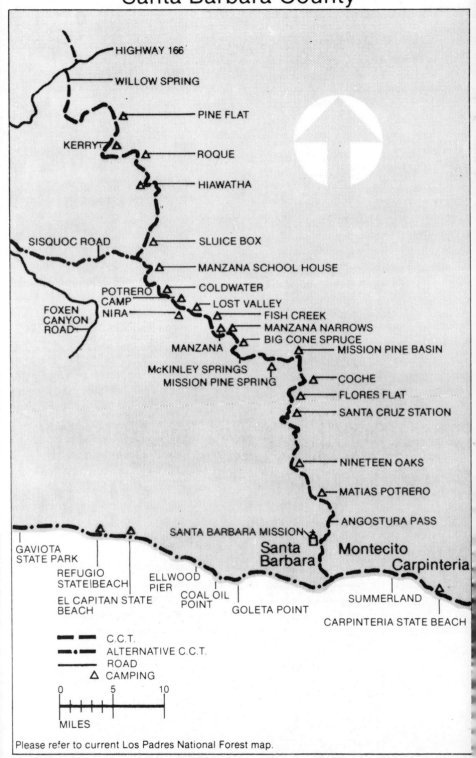

Santa Barbara County

From Carpinteria west, the Santa Barbara County shoreline extends to Point Conception, one sandy and mellow beach after another. The coastline's southern exposure results in clearer water, smoother sand, warmer sun. The beach is lined with a narrow coastal terrace that seems to protect the beach from residences, traffic, the hustle and bustle of the modern world.

Most of the north county coastline from Gaviota to the San Luis Obispo County line is inaccessible due to the presence of Vandenburg Air Force Base and the Hollister Ranch. Sea cliffs line this shore, seldom visited because of its distance from Highways 1 and 101. Still, there are some hidden beach treasures at Jalama Beach, Point Sal and Guadalupe Dunes awaiting those willing to venture off the beaten track.

Beneath the sharp peaks of the Santa Ynez Mountains is the city of Santa Barbara with its numerous red-tiled Spanish buildings and historic Mission. The city and its palm-lined beaches draw visitors from all over the world and each year becomes more of an international resort.

CCT travels first north then west into the mountains behind Santa Barbara, making a long sojourn into the Los Padres National Forest. About 12 miles, as the condor flies, behind Santa Barbara is the rugged San Rafael Wilderness. CCT ventures up chaparral-cloaked slopes, crosses oak potreros and climbs to pine and fir-covered peaks for breathtaking views of the Channel Islands. CCT winds along with Manzana Creek and the Sisquoc River through the wilderness, then follows the Santa Maria River west to the ocean at the Guadalupe Dunes.

Santa Barbara county offers a beautiful and diverse hiking experience— beaches and sand dunes, a resort town and a wilderness retreat. CCT hikers will find some of the easiest trail walking of their trek along Santa Barbara's sandy beaches and some experience of the most difficult paths in the San Rafael Wilderness.

> **CCT at a glance**
>
> **TERRAIN:** sandy beach from Carpinteria to Gaviota State Beach. Steep trails into brush-covered Santa Ynez Mountains; most follow stream drainages. Oak potreros and pine ridges. Travel along Sisquoc River rocky and wet. Swift travel possible atop graded banks of Santa Maria River.
>
> **OBSTRUCTIONS:** A beach route is thwarted by the Hollister Ranch up-coast from Gaviota and Vandenburg Air Force Base. CCT exits Los Padres National Forest along Sisquoc River, which passes through private property, the Sisquoc Rancho. Possible hassles. Parts of the Los Padres National Forest (and CCT) are closed to entry during the fire season, approximately July 1 to mid-October. A wilderness permit is required for travel in the San Rafael Wilderness.
>
> **TRANSPORTATION:** Amtrak stops in Santa Barbara. Transit info: Santa Barbara Metropolitan Transit District (MTD) P.O. Box 550, Santa Barbara 93120, (805) 962-7682.
>
> **CAMPGROUNDS/ACCOMMODATIONS**
> Rancho Guadalupe County Park
> Gaviota State Park
> Refugio State Beach
> El Capitan State Beach
> Carpinteria State Beach
> Plenty of trail camps in Los Padres backcountry along CCT. Motels, bed-'n'-breakfast establishments very expensive in Santa Barbara.

CCT: SANTA BARBARA COUNTY

After rounding Rincon Point, CCT enters Santa Barbara County. On the bluff above is Rincon Beach County Park a picnic area. In 1974, when beachgoers heard the park was slated to receive a heavy dose of cement paving and other "improvements," they staged a sit-in to protest. The powers that be were persuaded to pave less and preserve more.

CCT parallels the Southern Pacific tracks, quite close to the beach along this stretch, past the tiny community of Wave. Hikers may join the debate over whether Wave was named for its proximity to the ocean or because people "waved" at the train going by.

On narrow sandy beach, CCT enters the city limits of Carpinteria. On August 17, 1769, the Portola expedition observed Indians building a canoe and dubbed the location with the Spanish name for carpenter shop. You pass by some low dunes which border Carpinteria State Beach on the east. The state beach has a large (347 sites) campground and a special hiker-biker campsite. Carpinteria residents boast they have "the safest beach in the world," because although the surf can be large, it breaks far out and there's no undertow. As early as 1920, visitors reported "the Hawaiian diversion of surf-board riding."

Carpinteria Beach

"Rode along the beach a few miles today through the fog, visited some rocks of interest, frightened a large seal off from some rocks, saw thousands of gulls and geese, herons and other sea birds. After dinner at four o'clock, took a fine bath in the surf." —William Brewer, Carpinteria, Caliifornia March 5, 1860
Up and Down California: The Journal of William Brewer

The Carpinteria Tar Pits once bubbled up near the state beach. Spanish explorers noted that the Indians calked their canoes with the asphaltum. Around 1915 crews mined the tar, which was used to pave the coast highway in Santa Barbara County. In order to dig the tar, workmen had to heat their shovels in the furnace; the smoking tar would slice like butter with the hot blade. The tar pits trapped mastodons, saber-toothed tigers and other prehistoric animals. Unfortunately the pits, which may have yielded amazing fossils like L.A.'s La Brea Tar Pits, became a municipal dump.

CCT crosses the two-mile-long state beach to narrow, sandy Carpinteria City Beach. The beach narrows further as you approach the residential strip known as Sandyland. At high tide you may need to hike atop the sea wall. Ahead is El Estero, a tidal wetland separated from the ocean by a sandbar. An offshore shoal alters currents so that they drop much of the sand they carry at this place. Storm waves pile the sand above the normal tide line, forming the bar.

Author Stewart Edward White, who wrote about the California wilderness in such novels as *Fire, Storm,* and *Folded Hills,* purchased property here in 1910. A seaside resort was envisioned and wealthy Montecito residents built elegant houses along the waterfront from one end of El Estero to another. All went well until yeast magnate Max Fleischman furnished money to build the Santa Barbara breakwater. The ocean current south of Santa Barbara was altered and many homes and much coast washed away. Extensive sea walls have been built on both sides of El Estero.

During times of low water, you can wade across El Estero to Sand Point. If crossing is possible you'll continue along a narrow beach beside another breakwater to the brief commercial strip known as Santa Claus.

If El Estero is not fordable, backtrack along the breakwater a few hundred yards to Sandyland Cove Road. (You can try walking through the wetland, but there are many meandering streams and you're bound to get wet and mucky.) Sandyland Road crosses El Estero on a levee along the wetland's east border. Follow a trail along the Southern Pacific railroad tracks for one mile and rejoin the beach at the community of Santa Claus. While passing through the wetland, note the fine specimens of pampas grass with their tall plumes. At one time ranchers raised the plums, dried them, and shipped them east for sale. They were particularly popular in New Orleans, as decorations for Mardi Gras. A bit down-coast from Santa Claus, you may experience some difficulty getting around a small point with a white elephantine estate squatting atop it.

Both the beach route and the El Estero route converge at the Santa Claus business strip. A giant Santa stands atop a restaurant. Iron posts and rocks form a breakwater along the shore; there's not much beach and you'll probably have to rock-hop near a few houses and numerous private property signs. Try to stay on the shoreline, because once you retreat inland, it's difficult getting back to the beach, owing to a wall of homes and a lack of accessways. Keep going, alternating between rocky shores and patches of sand beach.

CCT rounds Loon Point past the mouth of Toro Canyon Creek. The point is believed to be of the same geologic origin as Rincon Point. Beyond this point, the beach widens and handsome bluffs rise above a sandy strip, favored by *au naturel* bathers.

CCT enters the town limits of Summerland on sandy beach. The beach here is sometimes beautifully cusped. Oil platforms line the horizon. Summerland was established a century ago as a tract of land, originally part of the pueblo lands of Santa Barbara. Later it was colonized by those with Spiritualist leanings and the name is taken from Spiritualist literature popular at the time.

In the waters here, the first offshore oil platform in the Western hemisphere was erected in 1896. Soon over three hundred wells were pumping oil from Pleistocene rocks at depths of 100 to 800 feet, an insignificant depth by today's standards. Commercial kelp harvesting also occurs here. Kelp produces chloride of potassium, a vital chemical in the production of gunpowder and other high explosives. When World War I broke out in Europe, the main source of that chemical was shut off to the U.S. so the Summerland Kelp Mill was established.

Above Summerland Beach is Lookout County park, a nice place to picnic. A well-marked ramp leads to the beach. From Lookout (Summerland) Beach, a sea wall extends ¾ mile to Fernald Point. At high tide it's possible to hike atop it, but you'll have to battle some brush. CCT passes a pretty little cove, bounded on the far side by Fernald Point, which was formed by a fan delta deposited at the mouth of Romero Creek.

Around the point, as CCT approaches Montecito, you'll see the higher parts of the Santa Ynez Mountains on the north skyline and the overturned beds of sandstone located near the peaks. There are no official public beaches in Montecito, but most of the shoreline receives public use. Fernald-Sharks Cove is one of three proposed public beaches. CCT soon crosses this narrow beach to Miramar Beach, below the Miramar Hotel. "Miramar-By-the-Sea" has been a popular watering place since the completion of the Southern Pacific railroad line in 1901. The hotel, with its finely landscaped ground and blue-roofed bungalows used to be a passenger stop. The Hotel's management is extremely private-property conscious and has actually had non-guest sunbathers arrested on "their beach."

In another quarter mile, CCT crosses Montecito's third beach, Hammond's, popular with surfers. Hammond's Meadows on the bluffs above the beach is a former Chumash habitation and a potentially rich archeological dig. It was recently placed on the National Register of Historic Places. George Hammond took off from a short air strip on the bluffs, flying his plane, loaded with mail and supplies, to the Channel Islands in the years before World War II. The remains of his hangar are still there.

Up-coast from Hammond's, CCT passes a number of fine homes. You are soon overwhelmed by the elegant Biltmore Hotel, built in 1927. CCT follows the hotel's narrow beach. In a short distance, the bluffs rise high above you and the shoreline grows rocky. Above is the Santa Barbara Cemetery. The beach becomes wide and sandy once more at Santa Barbara's East Beach. Across Cabrillo Boulevard is the Andree Clark Bird Refuge, where an enclosed salt water marsh provides habitat for ducks, geese, herons, and many native and migratory waterfowl. Adjacent to the bird refuge is A Child's Estate Zoological Gardens, a beautifully landscaped park and zoo especially designed for children.

Paralleling sandy Cabrillo Beach is beautiful Palm Park, as fine a collection of palms as you'll see on the west coast. The palms stretch for a mile along the shore to Stearn's Wharf. Until John Peck Stearns built his wharf in 1882, Santa Barbara's growth was slow because the city was geographically isolated by the mountains and the sea. The wharf established Santa Barbara as a seaport, offering mariners the longest wharf between San Francisco and San Pedro. Passenger steamships loaded with tourists tied up at the wharf. In the 1930s, a gambling ship, believed to be owned by gangsters, anchored just beyond the 3-mile limit. Water taxis brought gamblers from the wharf to the ship. In the 1940s the wharf was owned briefly by film star James Cagney and his two brothers, who planned to remodel it. They decided it was too expensive. The wharf, now owned by the city, was remodeled and reopened in 1981, and offers restaurants, shops and a fish market. Fishermen drop their lines from the end of the wharf, where they enjoy great sunset views of the Channel Islands.

State Street, Santa Barbara's main street, begins at the foot of the wharf. The next street up-coast is Chapala. The train station is two blocks inland; the bus station ten blocks inland. Since there's no campground or hostel in or near the city, you'll have to find accommodations in one of the many motels along Cabrillo Boulevard or at one of the Bed & Breakfast Inns.

From Stearn's Wharf, CCT heads inland on State Street. You'll cross Highway 101 and pass shops, restaurants, historic De la Guerra Plaza, and El Presidio. When you reach Mission Street, about twenty blocks from the ocean, turn right, following the street to its end, then bearing left 2 blocks to Mission Santa Barbara, often called "The Queen of Missions" and the tenth of the California missions to be founded by the Franciscans. It was established on the Feast of Santa Barbara, Dec. 4, 1786. The Mission and its gardens are well worth a visit. (You can take the #22 bus from downtown Santa Barbara to the old mission.)

CCT continues past the Mission on Mission Canyon Road. Almost immediately you'll pass Rocky Nook Park.

(An alternate route from Mission Santa Barbara to the backcountry trailhead is to follow the pathway through Rocky Nook Park for approximately ½ mile then connect with Mission Oaks Lane for ¼ mile before meeting Foothill Road.)

A turnoff to the left extends an invitation to visit Santa Barbara's Museum of Natural History, with excellent displays on the Chumash Indians and Southern California flora and fauna. In ½ mile turn right on Foothill Road, almost immediately turning left back onto Mission Canyon Road. Shortly, there will be a V-junction; veer left on Tunnel Road, following it 1.5 miles to its end at a locked gate. (A right at the V-intersection will take you to the Santa Barbara Botanical Gardens, an excellent place to learn about the state's native biotic communities.)

Make sure you have plenty of water before beginning the ascent out of Santa Barbara, particularly between May and November. Also remember that some parts of the Santa Barbara backcountry are closed during fire season and some stretches of CCT pass through these areas. Plan your hike accordingly.

Beyond the locked gate, CCT follows a crumbling paved rod, entering the Los Padres National Forest. After climbing one mile you reach a bridge, crossing seasonal Mission Creek flowing from the west. Cross the bridge, passing a waterworks facility belonging to the city of Santa Barbara. In times of abundant rain, the creek has some fine pools. This is where the tunnel from Gibraltar Dam Lake exits the Santa Ynez Mountains to supply Santa Barbara with water. After crossing the bridge stay to the left as the road becomes dirt and walk 1/8 mile under some handsome oaks to a junction with a California Riding and Hiking Trail sign.

The sign indicates the trail to Inspiration Point. A sharp right steeply uphill is the route of CCT (the Tunnel Trail) to Camino Cielo Road. No shade or water; however, spectacular views reward the hikers during the climb east toward La Cumbre Peak. In 1/2 mile the trail crosses the Southern California Edison Catway. The trail resumes immediately on the other side of the catway and rises above the powerlines. CCT switchbacks up the ridge for a mile, traverses a knoll, and in about 1/4 mile meets the junction with the signed Rattlesnake Trail which leads off to Gibraltar Road and Rattlesnake Canyon. At this junction continue on the Tunnel Trail, bearing left (north). The sign reads 2 miles to Camino Cielo. You follow the contours of a ridge for 1 1/2 miles. The last 1/2 mile of trail levels out, as it meanders through a small canyon to its intersection with paved East Camino Cielo in the saddle near Angostura Pass. If you want to bag La Cumbre Peak (3985'), it's a 3/4 mile hike west along the road.

At East Camino Cielo, bear left, then turn right shortly at Forest Road 5N25. You'll descend toward the Santa Ynez River, soon passing a locked gate. One quarter mile past the gate, keep a sharp lookout for the unmarked Tunnel Trail leading off to the left. You must join this steeply descending trail for more than a mile, sometimes only 150 yards away, the chaparral is impenetrable, making cross-country travel out of the question.)

A 3/4 mile descent brings you to the end of the Tunnel Trail and a junction with the Devil's Canyon Trail, which heads steeply down to the Santa Ynez River (the most direct route to water if you're thirsty). CCT continues left on the Matias Potrero Trail to Matias Potrero (1600') a small north slope meadow camp with two stoves and a table. Water is seasonal from a nearby creek. The camp is located at the former homesite of Matias Reyes, an old Indian who used to cut and sell wood in Santa Barbara.

Leaving the portrero, we hike first west then north, soon intersecting a signed junction pointing to the Santa Ynez River campgrounds. We leave the Matias Potrero Trail and descend to the north toward the Santa Ynez River on a jeep road (Matias Potrero Connector Trail). This trail intersects Live Oak Picnic Area. (Santa Ynez Campground is .2 mile to the right.) CCT bears left on the road for .1 mile. Hikers will look sharply for a small metal Forest Service sign indicating TRAIL off to the right (by the river). A short path leads from the road down to the Santa Ynez River, Ford the river (a difficult crossing in rainy season) at a wide spot and look for a sign on the other side of the river which reads: Camuesa Connector Trail, Hidden Potrero 4 miles.

NOTE: During dry season, areas north of the river are closed to hiking.

The trail ascends steeply through shadeless chaparral, giving increasingly good views of the Santa Ynez Mountains and River Valley. In 3 miles, just past a rolling nmeadow, Hidden Potrero, you'll intersect Camuesa Motorcycle Road. A one mile hike to the right on Camuesa Road brings you to Hidden Potrero Camp (2700′) located in a rolling meadow surrounded by oak woodland and chaparral. It was established as a base camp during construction of the Camuesa Road in 1927-1928. Water is piped from a year-round spring into water tanks at the upper end of the camp, which has a table and stove.

CCT goes right on Camuesa Motorcycle Road, usually quiet on weekdays, but buzzing with two-wheeled locusts on weekends. Look for an old mining road on your left with a cable across it. The road drops down to a flat area. Look for a trail branching off to the left and follow it down to Nineteen Oaks Camp. Oaks shade this camp, but not nineteen of them. The camp has a few tables and stoves. Geologically-minded hikers will note the distinctive outcroppings of Franciscan strata and scars in the nearby hills where mercury, also known as cinnabar or quicksilver, was mined.

CCT follows a spur trail from camp to the Santa Cruz Trail, heads north, and soon crosses Oso Creek. (Fill your canteens here; there's no reliable water for 4-9 miles.) The trail begins switchbacking through grassy meadows. Dipping in and out of oak and poison oak-smothered canyons, hikers ascend a hill to a saddle between the ridge we're traveling and Little

Pine Mountain. Switchback north, then west across the south face of Little Pine Mountain. The trail is steep and strenuous, but well constructed; it's chiseled into rock in places and stabilized by metal stakes and railroad ties. (No place for acrophobics!) In approximately 2½ miles you'll cross two large meadows, which in spring are smothered with purple lupine, poppies, and Indian paint brush. They almost look landscaped. For most of the year though, these meadows are tall dry grass—the habitat of deer and even a possible mountain lion!

The trail climbs around the heads of half a dozen canyons before reaching Alexander Saddle. To the left a bulldozed road goes to Alexander Peak (4107'). To your right a trail leads to Happy Hollow Camp (4300'). CCT heads straight ahead downhill along the signed Santa Cruz Trail. In a mile, just after passing a second connector trail leading to Happy Hollow Camp, you'll reach the junction with a short steep spur trail leading down to Little Pine Spring.

To reach Happy Hollow Camp from Alexander Saddle: Follow the right hand trail or make your own route over the ridgeline to the camp. Weather-worn pines on the ridgeline offer shade and you will catch fabulous views of the Channel Islands, Santa Ynez Valley, and Lake Cachuma. Happy Hollow, nestled among ponderosa pine, fir, and oak, has a few tables and stoves. Water is available at Little Pine Spring (see preceding and following paragraph). In the 1930's this camp was a recreation site for Civilian Conservation Corps workmen, who constructed the Buckhorn Road. A handsome field station, resembling a chalet, stood here until razed in the mid-1970s. The name Happy Hollow is apt; the camp is indeed in a hollow, and hikers are no doubt happy after a 3200' elevation gain in 6 miles. What hikers may not be so happy about in this north face camp is the lack of sunlight. The hollow traps the cold. Snow is often found here weeks after a storm, long after it has melted elsewhere.

From Happy Hollow, you may return to CCT (The Santa Cruz Trail) the way you came or take a shorter steeper connector trail. This trail leaves the hollow from West Big Pine Mountain Road on the northwestern edge of the campground. Follow the road 50 yards to the signed junction for the trail leading to Little Pine Spring. This trail ascends steeply through pines then descends 1 mile, rejoining the Santa Cruz Trail near the turnoff to Little Pine Spring.

The Little Pine Spring Trail, primitive and steep, leads down canyon to the left, and losing about 200 feet of elevation before reaching the spring. Here you'll find a seldom used camp with one stove and one table.

From the junction with the Little Pine Spring Trail, the Santa Cruz Trail (CCT) descends a sandy, wooded wash and across sometimes-swampy

meadowland. We then take a sharp turn and go northwest for about 2 miles on a cliff-hanger trail, known locally as "The Forty-Mile Stretch." Halfway along the stretch, stop and shout across the canyon; there's a wonderful echo. At the end of a canyon the trail turns hard right (north) and switchbacks steeply down to Santa Cruz Camp, one of the major entry stations for people hiking into the San Rafael Wilderness. Permits are required.

Santa Cruz Camp (1900'), situated on a meadow on the banks of Santa Cruz Creek, has excellent fishing and swimming areas, a half dozen stoves, tables, and a U.S. Forest Service Station and corral. Santa Cruz Creek runs year-round. At one campsite you can see the fireplace remains of a cabin built by F.W. Alexander, former owner of Rancho Oso. Ranchers ran cattle down the Alexander Trail, leading from Santa Ynez River to Santa Cruz Creek and the backcountry potreros.

Leaving Santa Cruz camp, the trail heads through meadowland near the ranger station northwest along the ridge above Black Mountain Creek through a mile of rolling meadows, the Roma Potreros, named after an early pioneer family from Goleta. You'll pass the junction with the McKinley Saddle. A sign welcomes you to the San Rafael Wilderness. Faint trail 28W07 to the left proceeds along the boundary of the Wilderness Area to Santa Cruz Peak. CCT takes the right fork, heading due north down to the west fork of Santa Cruz Creek. About ½ mile later we reach the creek and cross, following the east bank and entering Flores Flat, three miles from Santa Cruz Camp.

Flores Flat Camp, situated on the west fork of the Santa Cruz, offers campers fine views of nearby potreros. Two tables, two stoves. Leo Flores was a turn-of-the-century Mexican farmer, who built a cabin here, raised cattle and grew corn and watermelons. He irrigated his fields by digging an ingenious system of canals, which carried water to his fields from the creek. Flores even fished stray trout from his canals.

CCT meanders through Flores' meadowland, staying on the east side of the creek. The trail then veers eastward, climbing to Coche Creek, which hikers cross and re-cross several times. The trail leads along the southeast side of Coche Creek, passing the intersection with the Grape Vine Trail.

One-quarter mile past this intersection we reach Coche Camp (3320'), offering two stoves, a table, dependable water from the creek.

Coche means "pig" in Spanish and as the story goes, pigs escaped from ranches along the Santa Ynez River or were turned loose when hog prices were lower than the corn that fed them, and took up feral lives near Coche and Santa Cruz creeks. Boar hunters still bag some of the wandering swine.

Leaving Coche Camp, you hike along the east side of the creek for a while, then the west, fording Coche Creek one last time and beginning a long ascent. (This is the last dependable water until Mission Pine Springs, so take advantage.) The hiker gains over two thousand feet in the next few miles, making this one of the steepest and most strenuous climbs along the CCT in this neck of the woods. You head west, then north, trekking up switchbacks through dense, exposed slopes. The prolific brush has all but overgrown the poorly maintained trail in places. As you near Mission Pine Basin, the trail follows sporadic ducts over rocky outcroppings. Generally stay to your left, but choose your route carefully because many runoffs look like the main trail.

Mission Pine Basin normally receives snow in winter. No dependable water is in the vicinity, but two creek beds occasionally run. The water collects in pools; it should be treated. Mission Pine Basin was established as a trail maintenance camp in 1924. (Some hikers joke the trails haven't been maintained since.) A campsite is located beneath a lone pine tree in the center of the meadow by the Mission Pine Trail Junction. Beyond this junction the unmaintained trail continues along a ducted route by the creek bed to the main Mission Pine Basin Camp, where you'll find several stoves and tables under pine trees.

From the signed junction, turn west onto the Mission Pine Trail (28W01), following it through mixed chaparral and pine for 4 miles to Mission Pine Springs. The word "Mission" is prevalent in these parts because legend has that Jeffrey and sugar pine were logged from the forested sloeps to construct the Santa Barbara Mission in 1786 and Santa Ynez Mission in 1804. (One wonders how the padres dragged the trees down the mountain.) Good views are yours as you hike past castellated sandstone. Although you gain only 440' from Basin to Springs, you do so several times.

Mission Pine Springs is remote, secluded, beautiful with two tables, two stoves, dependable water. Homesteaders on the Sisquoc used to come here to cut shakes.

Continue out of camp on ducted trail. Ascending through pines and boulders, the trail loses its tentative feeling. In a little over a mile you leave the pines behind. An unsigned, 100-foot spur trail on your left takes you to San Rafael Peak (elevation 6,593'), the second highest peak in Santa Barbara County.

The trail becomes a fire road as you exit the San Rafael Wilderness. A very short spur trail to your right leads to McKinley Springs (5575′), a little known trail camp not shown on most maps. Around the spring is black oak, bay, chaparral. A table and stove complete this camp on the north slope of McKinley Mountain. During World War II, an aircraft observation station was operated atop McKinley Mountain. The spring, originally called Cold Spring, was renamed after it became the station's water supply.

Passing the big green water tanks on our right, CCT sticks with the fire road, which dips then begins to bend to the west. Here you'll find the unsigned junction with the trail leading down to Big Cone Spruce, which is the way CCT hikers will head. This trail is not a trail; the Forest Service intended to build one but never got around to actually constructing it. The Big Cone Spruce Connector Trail is actually a bulldozer line that was pushed down the side of the mountain during the Wellman Fire.

Before discussing CCT's forward progress down the difficult Big Cone Spruce Connector Trail, a way to exit the Wilderness at Cachuma Saddle Station on Sunset Valley Road is presented below:

Continue on the steep and winding road, descending around, but well above, the headwaters of Manzana Creek. We ascend briefly to reach Hell's Half Acre, an area of interesting rock formations. The road winds down around Cachuma Mountain to Cachuma Saddle Forest Service Station. The station is usually closed, except during fire alerts. There's parking here for wilderness-bound hikers.

The Big Cone Spruce Connector Trail is quite overgrown in spots, particularly at the beginning. The very steep three mile descent has been flagged by the California Coastal Trails Foundation.

CCT arrives at Big Cone Spruce Camp (3920′) just down canyon from the confluence of three creeks which make up the headwaters of Manzana Creek. Two stoves and a table beneath stately Big Cone Spruce. CCT heads down creek for 2½ miles along traces of washed-out trail. After plenty of creek crossing and boulder hopping, the unmaintained trail arrives at the junction with the Manzana Trail (30W13). CCT follows the Manzana Trail downstream ½ mile to oak and willow-shaded Manzana Narrows Camp (2960'). Just below the camp, which lies in a narrow part of the canyon, are some fine pools for fishing or cooling off.

Manzana means "apple" in Spanish and it's guessed an apple orchard once grew in the area. On second guess, it takes its name from manzanita, "little apples" in Spanish, a chaparral bush which produces a berry enjoyed by bears, foxes, and coyotes.

Leaving camp, a gentle ¾ mile descent brings you to Manzana, another canyon camp with two stoves and fine swimming holes. CCT continues down stream past tall thin alders and in season, wildflowers. The trail stays close to the creek and the canyon begins to widen quite a bit. Shortly, the trail crosses to the north side and proceeds above the creek to avoid the washouts below. In addition to a few stocked trout, which survive the legions of fishermen, you'll find frogs, crayfish and turtles in the Manzana. Deer and sometimes even bear prints are found in the sand at creek's edge.

After a mile, look to your left across the creek and you'll see Fish Creek Camp (2,000') on the far side of the Manzana flood plain, where Fish Creek Canyon meets Manzana Creek. Fishermen like this camp because the creeks here usually supports a large trout population. The somewhat overused camp, with two stoves, is set in an exposed area.

CCT continues its rolling, but gentle down creek descent about 1½ miles to Lost Valley Trail junction. A trail camp, with two stoves, is set amongst oaks and digger pine. The camp resides near the mouth of Lost Valley Canyon, which descends from Hurricane Deck to Manzana Creek. Hurricane Deck is a marvelous place to explore for the adventurous.

Leaving Lost Valley Camp, CCT continues its easy descent, switchbacking along and heading west along the creek's north bank. In 1 mile you reach NIRA Camp. The major entry point and our departure point for the San Rafael Wilderness, NIRA is an acronym for the National Industrial Recovery Act, a federal government program launched during the Depression. NIRA is an auto camp, heavily used, water from creek in season.

From NIRA, follow Sunset Valley Road back across the creek, turning right and continuing on the Manzana Trail and re-entering the San Rafael Wilderness. CCT heads northwest on good trail high above the creek. Your path switchbacks down to the creek crossing and the junction with the Potrero Canyon Trail and Potrero Camp (1600'). The camp offers two stoves and water from the creek.

CCT continues down creek on a gentle descent 2 miles to Coldwater Camp (1600') in a meadow seasonally smothered by lupine and shooting stars and surrounded by oak and pine. Water is dependable here even in summer. The trail pushes on down creek along the border of the Wilderness for five more miles, passing some excellent swimming holes. The trail leaves the creek for brief periods, traversing wildflower-dotted meadows. You'll pass the Dabney Cabin, built in 1914 by the Dabney family for use as a hunting lodge.

Manzana School House Camp (1150′) is situated on a flat area at the mouth of a canyon where Manzana Creek meets the Sisquoc River. Several tables and stoves. Water can be a problem in summer. Also, livestock from the Sisquoc Ranch graze this area and have severely polluted the water.

The old school house, now a county landmark, was constructed of lumber chopped and sawn from nearby stands of digger pine. It was erected in 1894 by eleven families who were followers of Hiram Preservid Wheat from Wisconsin, who, it was said, had the power to heal with his hands. Hostile Indians were so impressed by the spiritual power of this white man that they inscribed his wagon with a sign indicating he was to be granted protection. Today, the historic school house offers emergency shelter in bad weather.

Leaving Manzana Camp, CCT follows the Sisquoc Trail (30W12), actually a rough jeep road along the left side of the Sisquoc River. On the high banks above the riverbed is Wheat Mesa and the remains of a homestead belonging to Wheat's son-in-law Harly Wells. He owned the east mesa, his father-in-law the west; 1¾ miles of hiking brings you to a junction with the Horse Gulch Trail.

CCT hikers unfortunately are at something of an impasse at this junction. One trail, the Sisquoc-to-the-Sea Trail, leads through private property; the other, the continuation of CCT, is all-but-impassable because of brush. Both routes are described below.

CCT: SISQUOC-TO-THE-SEA TRAIL

The sign at Horse Gulch Trail Junction gives the mileage to the Goodchild Ranch. *WARNING:* You will soon be leaving National Forest land and entering the private property of the Sisquoc Ranch. As of this writing, the California Coastal Trails Foundation has not secured access to the ranch road which parallels the Sisquoc River for ten miles to Foxen Canyon Road. It would be highly desirable to open this road to foot traffic, because it's a logical entry point into the San Rafael Wilderness. Permission to walk the ranch road may be requested by writing: Supt. Harold G. Peiffer, Sisquoc Rancho, Rt. 1, Box 147, Santa Maria, CA 93454.

If you decide to hike down Sisquoc Ranch Road, please respect all structures. No camping or fires.

You continue down the road, passing an old chimney and the remains of a cabin and arrive at the locked gate of the Sisquoc Ranch. A sign posted on the gate reads "No Thoroughfare." Those hikers assuming this sign refers to vehicles, continue over the gate, the last one you'll see for many miles. Lovely old oaks shade the ranch road. Soon you'll see the remains of a second chimney. You'll ford the river, the first of over a dozen such crossings. In about 2 miles the road passes a handsome cabin, the Tunnel home, used during grazing season by ranch hands. Various cowpokes have made it their custom to carve their initials near the entrance.

A short way beyond the cabin, a jeep road that ascends sharply up to Pickett Corral intersects our route. Ignore it and continue down to the river. The riverbends are wide, and in dry season some hikers prefer to leave the road and "cut corners by hiking a diagonal from one bank to the other." Not much life is evident along the river, particularly in summer; sun and rocks in the riverbed, cattle on the brown hills, clouds of flies and mosquitos.

Our descent is leisurely, about 100 feet a mile. The significant geographical landmark is Bee Rock, encountered on a tight bend of the river; it takes quite an imagination to conjure a bee hive from this lump of rock. You'll pass another jeep road on your right which leads to the Goodchild Ranch.

A few miles from road's end you'll ascend a small rise and gaze at lovely views of the lower Sisquoc, checkerboarded with patches of green, irrigated oases in this dry land. You arrive at another locked gate. Beyond it, you'll follow a dirt road with an alfalfa field on the left. Cut over to a well-graded dirt road on your right. A vineyard is on the right side of the road. Farther along the road you'll see two crosses marking graves. Proceed down the road, keeping the grapes on your right.

The last 1½ miles of Sisquoc Ranch Road is paved. A barn, a few houses and the Sisquoc Winery are on the left. Thirsty hikers will no doubt delight in the joys of winetasting here.

One mile beyond the winery, Sisquoc Road intersects Foxen Canyon Road. On the bluffs above this intersection is the historic Sisquoc Chapel and cemetery. The chapel, now encircled by a cyclone fence, was built in 1876 to serve the 65 pioneer families of the Sisquoc area.

Foxen Road is rich in history. When General John C. Fremont and the American Army invaded California and marched on Santa Barbara, the Californios planned to annhilate the Americans at Gaviota Pass. Legend has it, American sympathizer Benjamin Foxen led the General to San Marcos Pass, through which Fremont and his followers escaped to Santa Barbara. Though revisionist historians have disputed Foxen's role in this historical episode, it does make a nice legend.

Here hikers must decide whether to tackle the long 25-mile trek to the sea along the Sisquoc, then the Santa Maria River. Walking along the riverbank is not difficult; the terrain, in fact, is much like that of the coast—sand and rock, minus the view of course—but it can be hot and monotonous and there are no official campsites enroute. One alternative might be to catch a ride into Santa Maria and hike the last dozen or so miles to the ocean.

Those wishing to push on will return with CCT to the Sisquoc River, following its winding course past bean fields and oil wells. Vineyards are replacing more traditional cash crops as the California wine industry booms. In contrast to the peaceful vineyards, a sand and gravel operation is being conducted in the area and heavy trucks lumber by on riverfront roads. In a few miles you'll pass the company's base of operations.

The trail passes under Santa Maria Mesa Road Bridge. (If you follow Santa Maria Mesa Road ½ mile west, then make a left turn on Route 176, you'll see the Garey General Store, which sells welcome refreshments.)

A short distance past this bridge is Fugler Point, where the Santa Maria River is formed by the confluence of the Cuyama and Sisquoc Rivers. Except during a flood, it's not a dramatic meeting of the waters. From this point, the Santa Maria River drops about 400 feet over the next 20+ river miles.

The Santa Maria River Valley was once a bay, filled to the brim by an ancient sea. In Miocene times, gigantic toothed marine birds flew overhead and whales patrolled the waters; the fossils of these animals have been discovered in the flagstone of nearby hills. For eons, land rose and sea receded. The ancient bay floor eventually became dominated by two rivers that formed a third—the Santa Maria River, which rushed to the ocean during the wet season.

A surprisingly small amount of human activity is apparent along the river until you reach the outskirts of Santa Maria. A subdivision or two can be glimpsed in the distance. The sandy channel of the river has been building up for thousands of years. About 3 miles east of Santa Maria the river bed is actually 50 feet higher than the town. The trail passes under the small Bull Canyon Bridge and in about 1½ miles reaches the long bridge where Highway 101 spans the river. An RV campground is nestled in the shade of the freeway.

Hikers may wish to follow the levee on the south bank of the Santa Maria; it's lots easier going. From the levee you'll be able to see quite a few miles down into the Santa Maria Basin, one of the most fertile coastal drainage basins in California. You'll spot sugar beets, and a multitude of vegetables. Both small truck farms and huge corporate farms line the river.

Approximately 5 miles from Highway 101, you'll pass over the first of two roads crossing the river, Bonita School Road, traveled by all sorts of strange-looking (to the city-dweller) farm machinery. You may be passing now and then into San Luis Obispo County and not even know it, because the county line snakes from one side of the river to the other.

Another 4 miles of hiking brings you to the Highway 1 bridge. South of the bridge, fronting Highway 1, is the sleepy Mexican-American town of Guadalupe. Hikers may pause to sample what many consider some of the best Mexican food on the central coast at the small cafes. Le Roy County Park a day use site with a nice picnic ground is perched on the south riverbank near the highway bridge.

If river walking is wearing you down and you feel the old cemetery in Guadalupe may be your final resting place, follow Main Street out of town; it leads 5 miles to the Pacific, paralleling the Santa Maria River.

River travelers will journey through more agricultural land. The river channel broadens and its mouth opens to the sea. Soon sparkling white sand dunes (and oil wells) come into view. The river mouth itself can be something of a quagmire and there's some quicksand in places, so it might be advisable to follow Main Street the last mile or so to the dunes.

Rancho Guadalupe Dunes County Park just south of the mouth of the Santa Maria River is little more than a parking area for recreational vehicles. It's very windy here and hikers would be better off searching for a wind-free hollow in the dunes.

CCT TO HIGHWAY 166

CCT, when completed, will *not* return to the beach via the Santa Maria River. The trail will continue in a northerly direction through the Los Padres National Forest, traveling the length of Santa Barbara's backcountry, crossing Highway 166, passing into the Santa Lucia Wilderness, and returning to the coast at Morro Bay.

Because the Sisquoc and Santa Maria Rivers, as well as adjoining areas are private property, CCT must remain within the boundaries of the National Forest. In short, a trail connecting the Manzana/Sisquoc River area with the maintained Kerry Canyon Trail to the north and Highway 166 farther north is a missing link in the CCT. Fortunately, the link isn't entirely missing, just desperately overgrown with brush in places; it can be followed by skilled and experienced bushwhackers.

From the turn-of-the-century, foresters have been aware of the need for a Sisquoc-Kerry Canyon route and in 1909 construction began. Trails forming this route were maintained and used until the advent of World War II. However, since the war, the trails have received little or no maintenance.

Trail tread exists, but brush has severly overgrown the route; only bears frequent the trails these days.

CCT departs the Sisquoc River at the mouth of Horse Gulch and proceeds up-gulch until it climbs over into the upper part of the south fork of La Brea Canyon. After descending this canyon a short distance, it again rises up and over into the drainage of Roque Canyon and follows this canyon down to its junction with Kerry Canyon. Both the upper and lower stretches of Kerry Canyon Trail intersect the Kerry Canyon fork of La Brea Canyon fork of La Brea Canyon Road. From here, the hiker can follow maintained trail to Highway 166 and points north. Situated along this route are several backcountry camps with water flowing from nearby year-round creeks. From South to North they are: Sluice Box, Hiawatha, Roque, French, Kerry.

The current Los Padres National Forest map does not picture some of these trails; you'll need a 1972 or earlier map, or topos. Jim Blakley, chairman of the Santa Barbara County Riding and Hiking Trails Advisory Committee, and the California Coastal Trails Foundation suggest the following route for skilled and rugged bushwhackers. It has been flagged by the California Coastal Trails Foundation.

From the Sisquoc River, CCT follows the signed Horse Gulch Trail, heading up-gulch and crossing the creek several times. In 2¼ miles the trail arrives at Sluice Box Camp. Constructed in the 1930's, the camp was the site of a mining operation where an early homesteader sluiced for gold. The camp is popular with hunters during deer season.

The first third of the climb beyond Sluice Box up Horse Gulch is washed out and very difficult to locate because of brush. The upper two thirds of the trail ascending with the gulch is quite brushy, but traceable. Cresting a ridge the trail descends through oak grassland and is easy going. The trail crosses over to the south side of the south fork of La Brea Canyon down to Hiawatha Camp on washed-out dozer trail, crowded with poison oak and alder trees.

Leaving Hiawatha Camp, CCT ascends along a dry stream channel, choked with dense chaparral; a brutal route. When CCT descends to the bottom of Roque Canyon, it's easier going. The Sierra Club cut brush along this part of the trail several years ago, but gave up before reaching the top of the ridge. Descending Roque Canyon, the trail switchbacks to a steep yucca-covered shale slope. Atop the ridge, CCT follows a brush-choked dozer trail, then begins a switchbacking descent, also brush smothered, into Flores Canyon. The trail improves greatly as it follows the north side of Roque Canyon down to its junction with the maintained Kerry Canyon Trail.

CCT heads up-canyon (right) on the Kerry Canyon Trail along La Brea Creek. (A left turn on the trail brings you to Lazy Camp in 1 mile. One table, one stove. Another ¼ mile past Lazy is the terminus of La Brea Canyon road.) Kerry Canyon Trail stays on the canyon bottom for a mile up-canyon. The trail then ascends a considerable distance above the west bank, then down to the creek bottom, then up on the west bank, then down again. Three miles from the junction with Roque Canyon we reach Kerry Camp, consisting of two ice can stoves located in a cluster of live oak on the west bank of La Brea Creek.

Leaving camp, the trail continues back and forth across the creek. One half mile beyond camp observant hikers may spot an active bear trail leading up to a spring. Three quarters of a mile from Kerry Camp, the canyon forks and the trail ascends between the canyon branches on a hog-back ridge to a trail fork indicated by an iron pipe. (The right fork, traveled by cows, leads to a dead end.) The main trail bears left and soon passes an area of blackened earth, once a Chumash yucca-roasting site. The Indians baked and ate the yucca flower stalks, which resemble giant asparagus. In the creek bed, opposite this site, is a seep spring.

The trail leaves La Brea Creek. (A trail, now all but vanished, once followed the creek to the top of Miranda Pine Mountain.)

Hikers soon reach Forest Road 11N03. On the other side of the road is the now-abandoned Pine Flat Picnic Area. A sign indicates it's 2 miles to Miranda Pine Camp on Miranda Peak and 15 miles to the road's junction with Highway 166.

CCT continues with Kerry Canyon Trail across the road for a few miles to the unsigned junction with Indian Trail (31W02) and bears left down Pine Canyon, following this trail 3 miles to Brookshire Spring, an auto campground at the terminus of Forest Road 11N04A. (Warning: The Forest Service frequently changes the numbers on their roads; this one used to be 12N03.) Follow this road ¾ of a mile to the unsigned junction with Willow Spring Trail (31W01) and go right. The trail soon passes an unsigned junction with a no-longer-maintained side trail, which leads off to Porter Spring. Our trail continues past this junction, crosses Aliso Creek, and passes a private ranch.

Farther up the trail is another unsigned junction for another unmaintained side trail, Willow Connector Trail, leading off to Willow Springs (on private property). For the next few miles, the trail skirts private property along the Forest boundary before intersecting Highway 166 just below Garcin Ranch.

From Highway 166, you may catch a ride over to the coast and resume your hike. The California Coastal Trails Foundation is in the process of planning a route northward of the highway to the Santa Lucia Wilderness and points north.

Good luck.

Biltmore Pier (demolished in 1983)

Summerland Trail (SB-1)

Lookout County Park to Biltmore Beach; 5 miles RT

One might guess Summerland was named for the weather, but the name was taken from Spiritualist literature—something to do with the Second Heaven of Spiritualism. A century ago, Spiritualists pitched their tents on tiny lots in the area.

In the waters here, the first offshore oil platform in the Western hemisphere was erected in 1896. Oil attracted far more people to Summerland than Spiritualism and soon the town's air was heavy with the smell of gas and oil. It was said that free illumination came easy—one simply pounded a pipe into the ground till reaching natural gas, and lit a match. Liberty Hall, the Spiritualists' community center, glowed with divine light and for a time Summerland became known as the White City.

This hike travels due west along sandy Summerland Beach, parts of which are popular with the clothing-optional set, rounds some rocky points, and concludes at the narrow beach in front of the famed Biltmore Hotel.

Directions to trailhead: From Highway 101 in Summerland, turn seaward to Lookout County Park. Mass Transit: MTD Route 7 to Summerland Post Office. Walk south on Evans Road to Lookout County Park.

The Hike: At Lookout County Park is a picnic area and a monument commemorating the first offshore rig. A well-marked ramp leads to the beach. From Lookout (Summerland) Beach, a sea wall extends ¾ mile west to Fernald Point. At high tide, you may wish to hike atop it, but you'll have to battle some brush. You soon pass a pretty little cove, bounded on the far side by Fernald Point, formed by a fan delta deposited at the mouth of Romero Creek.

Around the point, as you approach Montecito, you'll see the higher parts of the Santa Ynez Mountains on the north skyline and the overturned beds of sandstone located near the peaks. There are no official public beaches in Montecito, but most of the shoreline receives public use. Fernald-Sharks Cove is one of three proposed public beaches. You cross this beach to Miramar Baech, below the Miramar Hotel. The Hotel's management is one of the more private property-conscious groups on the California coast and in past years has actually had non-guest sunbathers arrested on "their" beach.

In another ¼ mile, you'll begin hiking across Montecito's third beach, Hammonds, popular with surfers. Hammond's Meadows on the bluffs above the beach is a former Chumash habitation and a potentially rich archaeological dig. Although it was recently placed on the National Register of Historic Places, it's been bulldozed for development.

Up-coast from Hammond's you'll pass a number of fine homes, travel around storm-damaged Biltmore Pier, and arrive at narrow Biltmore Beach, frequented by the rich and beautiful. Opposite the beach is the magnificent Biltmore Hotel, built in 1927.

Return the same way.

Stearns Wharf

Santa Barbara County Beach (SB-20)

Santa Barbara to Gaviota State Beach; 35 miles one way

While CCT heads inland from Santa Barbara into the Santa Ynez Mountains, beach hikers may continue up-coast for several more miles. It's easy walking; however, it's a little over 20 miles to the first campground at El Capitan State Beach.

Directions to trailhead: Begin at Stearns Wharf at the foot of State Street in Santa Barbara.

The Hike: West of Stearns Wharf, the trail travels over appropriately-named West Beach, a wide sandy stretch extending to Santa Barbara harbor. The harbor is home to many pleasure craft and a small commercial fishing fleet. Hikers improvise their way through the harbor past the Santa Barbara Yacht Club at Point Castillo, returning to water's edge at wide, sandy Ledbetter Beach. Coastal hydro-dynamics experts have studied this area as a classic case in beach erosion. Santa Barbara and points down-coast have suffered beach erosion since the harbor was built in the 1920s. When sandy sediment that would normally move down-coast and replenish the beaches is trapped by the harbor breakwater, beaches down-coast suffer intense erosion and the harbor itself becomes clogged. Such sediment trapping makes frequent and expensive dredging operations necessary.

Beyond Ledbetter Beach the shoreline narrows again and passage is possible only at low tide. Cliffs grow higher as the trail rounds Santa Barbara Point, a popular surfing spot. On the tall bluffs above, Shoreline Park provides picnickers with bird's-eye views of the harbor, channel and islands beyond. A little farther on, from the tall bluffs above the beach, Santa Barbara Lighthouse flashes its welcomes and warnings to channel mariners.

A mile's hike over sandy and cobbled beach brings you to Arroyo Burro Beach county Park. At the turn of the century the Hendry family owned the beach and it was known as Hendry's. Today some Santa Barbarans refer to it as "Henry's." The beach was officially re-christened Arroyo Burro in 1947 when the state purchased it for $15,000. The park, later given to the county, was named for the creek which empties into the ocean at this point. Arroyo Burro is popular for picnicking, sunbathing and hang-glider watching.

Beyond Arroyo Burro, walking is easy on wide beach. On the high bluffs above is the exclusive community of Hope Ranch. Residents often ride their horses along the surf line.

More Mesa Beach follows, one of the most peaceful beaches in the county. Only the buzzing of innumerable flies disturbs the tranquility of the serious sunworshippers who bake their hides here.

Beyond More Mesa is another mile of sandy, kelp-strewn beach. The sea cliffs here and in other parts of Santa Barbara have receded 3 to 10 inches per year or roughly 50 feet per century. While this erosion is less than other parts of the world—the White Cliffs of Dover, for example—it's still substantial enough to be a consideration for builders of bluff-top houses.

In 1981, five ancient cannons were found along Goleta Beach, a half mile south of Goleta Slough. Pacific storm waves exhumed the cannons, which are believed to have been made in the 1700s and of British origin. Historians speculate the cannons are from British ships, which used to lurk along the coast waiting for Manila galleons the Spanish in the Phillipines used to send the treasure they collected.

Hikers soon arrive at the Goleta Slough, large tidal mud flats that lie between the UCSB campus and the Santa Barbara Airport. Atascadero Creek empties into the slough, where a great variety of birds, crustaceans, and native flora thrive. Goleta Landing was built in 1897 and served as a shipping point for Goleta Valley walnuts and citrus. Pioneer families enjoyed boating, fishing, duck hunting and camping on the slough. Cabrillo's storm-battered ships are said to have taken refuge here. Although the slough is smaller than it was before bulldozing and flood control projects, in the 1960s it was saved from a Santa Barbara mayor's pet plan: a speedboat lake surrounded by a racetrack for sports cars.

Hikers wade the shallow, sandy-bottomed slough, resume walking on sandy beach, and enter Goleta Beach County Park. Here is a fishing pier and a grassy picnic area along the beach.

Continue up-beach past handsome sandstone cliffs. If the tide is especially high, you may detour up the cliffs through the U.C.S.B. campus. A mile and a half from the county park you'll reach Goleta Point, a popular surfing spot and begin heading due west. You'll pass a nice tidepool area; judging from the number of students and copies of *Between Pacific Tides* it's well studied. A little up-coast is sandy Isla Vista Beach where students sunbathe—and sometimes study.

The trail rounds a second point, Coal Oil, and crosses dunes, stabilized with grasses and rushes. Behind the dunes is the Devereux Slough, a mixture of salt and freshwater providing a unique habitat for native and migratory waterfowl. Bird watchers sight snowy egret, great blue heron, black-bellied plover and western sandpiper.

Two miles farther on a sandy beach dotted with globs of tar from natural oil seepage, you come to an old barnacle-covered drilling platform at Ellwood Oil Field. On February 23, 1942, a Japanese submarine surfaced off Ellwood Beach and lobbed 16 shells at the oil facilities. Tokyo claimed this first (and only) attack on the U.S. mainland "a great military success," though the incredibly poor marksmen managed to inflict only $50 worth of property damage.

Ellwood Oil Field is still active and over the next few miles you'll hear the occasional chugging of oil well generators from the bluffs above. Between Ellwood Pier and Gaviota State Park, the Southern Pacific Railroad tracks hug the coast. Usually walking on this stretch of coast presents no problems, but if you should have difficulty at high tide, you may follow the tracks. An easement has been granted by Southern Pacific to Santa Barbara County to build a hiking-biking trail along the right-of-way, but trail construction has not yet begun. If you decide to travel along this inland route, observe private property and stay off trestles.

A few miles past Ellwood Pier is the mouth of Dos Pueblos (Two Villages) Canyon, now generally known as Naples on the map. Legend has it that during Cabrillo's time an Indian village existed on each side of the creek. The inhabitants of each village were of a different race and spoke different languages; those on one side were short, heavy-set and swarthy and on the other tall, slender, fair. So the story goes.... Some San Francisco capitalists purchased land on the east side of Dos Pueblos Creek and laid out a townsite which they named Naples. Many lots were sold during a land boom in 1887. CCT hikers who journeyed through Los Angeles County will recall the Naples development near Long Beach, which was considerably more successful. Not much remains of this Naples. The land in these parts is owned mostly by Dos Pueblos Ranch. At the beach is a tiny duck pond, a pleasant resting place for weary hikers.

Up-coast from Dos Pueblos, streams dissect the cliff face and the Santa Ynez Mountains march closer to the area. As the trail rounds Edwards

Point and pushes on toward the state beaches, you'll pass one of the largest railroad culverts, built across Las Llagos Canyon in 1899. The culvert provided one of the final links in the railroad connecting isolated Santa Barbara with San Francisco.

El Capitan State Beach, located at the mouth of Canon de Capitan is a narrow, rocky beach with 142 campsites. A private campground, El Capitan Ranch Park, is across the highway.

"Capitan" refers to Captain Jose Francisco de Ortega, a fat and jolly Spanish Army officer who served as a trail scout for the Portola expedition. When he retired from service to the Crown in 1795, he owed the army money and offered to square things by raising cattle. The land he chose was a coastal strip, two miles wide and 25 miles long extending from just east of Point Conception to Refugio Canyon. Alas, Captain Ortega's retirement was short-lived; he died three years later and was buried at the Santa Barbara Mission.

Beyond El Capitan, hikers travel along mixed sandy and rocky beach. Sea cliffs are steep here, because they are constantly being cut back by wave erosion at their base. You'll pass unusually large Corral Canyon, its walls covered with beds of highly deformed light-colored shales.

Approaching Refugio State Beach, you'll see abundant kelp just offshore. If a breeze is blowing over the water, note how areas with kelp are smooth and kelpless areas are rippled.

Refugio State Beach, at the mouth of Refugio Canyon, is a rocky beach with tidepools. It has 85 campsites and a special hiker-biker group camp.

After the death of El Capitan Ortega, the Ortega family continued lving in Refugio Canyon for many years. The pirate Hippolyte de Bouchard landed in 1818 and sacked and burned the adobe casa that stood at the canyon mouth. Today, a famous actor and modern-day rancher, Ronald Reagan, makes his home atop Refugio Canyon.

Head up-coast along mixed rocky-sandy shoreline—or if you prefer, a one-mile long bluff trail will take you up-coast past some pleasant picnic spots. The beach soon widens and grows more sandy. You will travel over nine miles of almost completely undeveloped shoreline on the way to Gaviota. A mile south of Gaviota State Park, along Alcatraz Beach, you'll encounter some great tidepools. The state park has vehicle campsites, tent campsites and a fishing pier. Gaviota State park extends across Highway 101, where you'll find a trail to a hot springs and to Gaviota Peak.

The beach trail ends here at Gaviota Beach. Beyond is Hollister Ranch, which denies public access at this time. While court battles over access rage on, be warned that trespassers are dealt with severely by armed deputies. We hope before too much more time elapses, hikers will be able to complete the length of Southern California to Point Conception.

Goleta Beach Trail (SB-3)

Goleta County Park to Coal Oil Point; 7 miles RT
Goleta County Park to Ellwood Oil Field; 12 miles RT

Around 7:00 on the evening of February 23, 1942, while most Californians were listening to President Roosevelt's fireside chat on the radio, strange explosions were heard near Goleta. A Japanese submarine surfaced and its gunners aimed 16 shells at Ellwood Oil Field and Southern Pacific railroad tracks. They missed. The submarine disappeared into the night, leaving behind air-raid sirens, a jumpy population, and lower Santa Barbara real estate values.

This hike along Goleta Beach to Ellwood Oil Field is interesting for more than historical reasons. On the way to the Oil Field/Battle Field, you'll pass tidepools, shifting sand dunes, and the Devereux Slough, a unique intertidal ecosystem protected for research purposes.

(For description of the coastline north and south of this hike, see Santa Barbara County Beach Trail.)

Directions to trailhead: From Highway 101 in Goleta, head south on Route 217 for two miles to Goleta Beach County Park. Park in the large lot.

The Hike: Walk up-coast. In ¼ mile you'll reach a stretch of land called the Main Campus Reserve Area, where you'll find Goleta Slough. The Slough, host to native and migratory waterfowl, is a remnant of a wetland that was once more extensive.

Continue up the beach past handsome sandstone cliffs. Occasionally a high tide may force you to detour onto the cliff through the university campus. A mile and a half from the county park, you'll round Goleta Point and head due west. You'll pass a good tidepool area. Two more miles of beach-combing brings you to Coal Oil Point. You'll want to explore the nature reserve here. (Please observe all posted warnings; this is a very fragile area.)

The dunes are the first component of the reserve encountered on the seaward side. Pick up the trail over the dunes on the east side of the reserve. The fennel-lined trail passes under cypress trees and climbs a bluff above the slough to a road on the reserve's perimeter. The road is a good observation point for the slough, a unique ecosystem. Something like an estuary, a slough has a mixture of fresh and salt water, but an estuary has a more stable mixture. The water gets quite salty at Devereux Slough, with little fresh water flushing.

Birdwatchers spot snowy egrets and great blue herons, black-bellied plovers and western sandpipers. Birdathons, marathon bird sighting competitions are often held at the slough.

Option: To Ellwood Oil Field. Return to the beach and continue hiking up-coast. Sometimes horses gallop over the dunes, suggesting Peter O'Toole and Omar Shariff's meeting in "Lawrence of Arabia,"...except there's oil on the beach, as you'll readily notice when you look at your feet. In two miles you'll pass under an old barnacle-covered oil drilling platform and enter Ellwood Oil Field. Here the Japanese fired shots heard 'round the world...and missed.

Seven Falls Trail (SB-4)

Tunnel Road to Seven Falls; 3 miles RT; 400′ gain

> "A pleasant party spent yesterday up Mission Canyon, visiting the noted Seven Falls and afterwards eating a tempting picnic dinner in a romantic spot on the creek's bank. To reach these falls requires some active climbing, able-bodied sliding and skillful swinging..."
>
> —Santa Barbara *Daily Press*, 1876

Seven Falls has been a popular destination of Santa Barbarans since before the turn of the century. The seven distinct little falls found in the bed of Mission Creek are still welcoming hikers. While the falls are all but dry in summer, water tumbles gracefully over the rocks in winter. In spring, wildflowers bloom between the canyon walls, framing a perfect picture.

This hike follows a dirt road, then a short section of the Jesusita Trail to Mission Creek. The falls are a short scramble up the creek.

Directions to trailhead: From the Santa Barbara Mission, drive up Mission Canyon Road, turning right a block on Foothill Road, then make an immediate left back onto Mission Canyon Road. At V-intersection, veer left onto Tunnel Road and drive to its end. Park just before the locked gate.

The Hike: Hike from the end of Tunnel Road. You leave the power lines behind and get increasingly grander views of Santa Barbara. You cross a creek over a bridge. It's just under a mile to the Jesusita Trail, which departs from a small grove of oaks above the tunnel keeper's home site. (The house burned down during the Coyote Fire.) The trail descends steeply to Mission Creek.

At the creek bottom, hike up-creek about ⅛ mile into a steep gorge cut into solid sandstone. Geologists will recognize fossilized layers of oyster beds from the Oligocene Epoch.

For thousands of winters, rainwater has rushed from the shoulder of La Cumbre Peak and cut away at the sandstone layers, forming seven distinct falls and several deep pools. The higher falls are difficult to climb.

Return the same way.

Jesusita Trail (SB-5)

Cater Water Filtration Plant to Inspiration Point; 6 miles RT; 1000′ gain
To Junction with Tunnel Trail 8 miles RT; 1000′ gain

This lovely hike takes you along wooded San Roque Creek. Jesusita Trail is part of the U.S. Forest Service trail system and in 1968 became part of the California Riding and Hiking Trail System. Inspiration Point offers sweeping views of Santa Barbara, the east-west coast, the Channel Islands and Goleta Valley. The trail provides a link with the Tunnel Trail, near the head of Jesusita Canyon. From here hikers can walk to the top of the front range of the Santa Ynez Mountains.

Although the trail has been used continuously for a century, a few years back a maniacal private property holder bulldozed parts of the trail and hung "No trespassing" signs of the canyon. He was stopped and the public's right-of-way affirmed. Avocado groves are sneaking down the slopes of the canyon, but most of the creek bottom remains in a natural state.

Directions to the trailhead: From Highway 101 in Santa Barbara, exit on Las Positas Road. Head inland. Las Positas becomes San Roque Road past Foothill Drive. The water infiltration plant is on your left and there's parking at the trailhead.

The Hike: The signed trail follows first the east side then the west side of the creek. The trail meanders up to a meadow then back down to the oak-shaded creek. (If the trail near the creek is washed out from winter floods, simply follow the creek or the dirt road above.)

Signs direct you past the pasture near Moreno Ranch. The trail resumes to the left. It climbs a side canyon for ½ mile then begins switchbacking up a small hill, levels out briefly, then reaches a firebreak road. Follow this road under transmission lines, rising above them toward Inspiration Point.

There are several inspiring points, actually, and if you bushwhack through the chaparral a bit you'll find some sandstone rocks upon which to sit and picnic and enjoy the sweeping views.

Option: To Tunnel Trail. Follow the trail as it contours to the east, providing continuing good views of the coast. Take switchbacks down to a small side creek, cross the creek, and in a few hundred yards reach the junction with the Tunnel Trail.

Gaviota Peak Trail (SB-6)

Highway 101 to Gaviota Peak; 11 miles RT

Father Juan Crespi of the 1769 Portola Expedition dubbed the coastline here San Luis in honor of the King of France. However, the soldiers of the expedition thought that La Gaviota, Spanish for "sea gull," was a more apt description and this name stuck. The peak took its name from the beach.

This hike begins in Gaviota State Park and ends in the Los Padres National Forest. You'll visit warm mineral pools, then continue to the top of Gaviota Peak (elevation 2,458 feet) for superb views of the Santa Barbara County coast. Although a trail is indicated on the 1978 Forest Service map, it has since been bulldozed into a fire road.

Directions to trailhead: 35 miles up-coast from Santa Barbara, exit 101 at the Lompoc/Highway 1 sign. Turn east a short distance, then go right .2 mile to a parking area.

The Hike: Follow the fire road, which is strewn with rocks and eroded from winter floods. This is a well used stretch of trail; most hike only to the hot springs and turn back. You'll soon leave the noise of the highway behind. A half-mile walk beneath spreading oaks up the moderate grade brings you to Gaviota Hot Springs.

Cross the Springs' rock dam, climb over a fallen tree, and follow the trail under the gnarled limbs of valley oaks. The trail grows fainter and steeper as it leads through tall grasses bespeckled with poppies, morning glory, and wild rosemary. A half-mile climb from the springs brings you to a fire road. Bear right.

Hike up the fairly steep fire road, leaving behind the oaks and enter a typical chaparral community. Keep your eye out for a short unmarked trail on your right, which leads to a hidden grotto with a tiny waterfall.

The trail levels out for awhile. It's blustery up here on this ridge route. At times you might be assaulted by clouds, which zoom past, enclosing you in fog for a few seconds.

On clear days the Channel Islands and Point Conception are visible from Gaviota Peak. You'll be able to distinguish the unbroken coastline south from the Point to Hollister Ranch, Refugio, El Capitan.

Return the same way.

Point Conception Trail (SB-7)

Jalama County Park to Pt. Conception; 12 miles RT

At Point Conception, the western-running shoreline of Southern California turns sharply northward and heralds a number of changes: a colder Pacific, foggier days, cooler air. Ecological differences between the north and south coasts are illustrated by the differing marine life occupying the two sections. Point Conception serves as a line of demarcation between species of polyps, abalone, crabs and limpets.

The Point was first discovered by Juan Rodriguez Cabrillo in 1542. Raging storms prevented Cabrillo's passage and he was forced to anchor his ships at San Miguel Island. He waited eight days, tried to sail again and was driven back, this time off Gaviota Pass. He sent his men ashore for wood and water and set sail again, this time successfully rounding the Point.

The name Punta de la Limpia Concepcion was applied to the point by a later explorer, Vizcaino. He arrived on December 8, the day of Purisima Concepcion (Immaculate Conception) 1602.

This day hike leaves from Jalama County Park, the only genuinely public access point between Gaviota State Park and Jalama County Park. This part of the coast is divided between two huge ranches, Bixby-Cojo and Hollister. Hikers are required to remain below mean high tide when "crossing" ranch land. Be warned, they take private property seriously around here.

Directions to trailhead: Jalama County Park is located 20 miles southwest of Lompoc off Highway 1. From Santa Barbara, go 35 miles north of Highway 101, exiting on Highway 12 north. Continue 14 miles to Jalama Road. Turn left and go 14 more miles through some beautiful ranch country to the county park. Park in the day use lot. There's a fee.

The Hike: Head south over the splendid sand dunes. As you walk along the coast, you'll soon realize that although Jalama Beach is not on the main L.A.-San Francisco thoroughfare, two groups have found it and claimed it as their own—surfers and surf fishermen. The beach takes its name from Jalama, a Chumash Indian Rancheria of La Purisima Mission established in 1791. The ranch land and beach have changed little in the last 200 years.

Jalama County Park includes only ½ mile of shoreline, so you soon hike beyond the park boundary. The sandy beach narrows and gives way to a rockier shore. Offshore, on the rock reefs, seals linger. Depending on the tide, you'll encounter a number of sea walls. The smooth tops of the sea walls make a good trail. "1934" is the date imbedded in the concrete walls.

A half mile from the lighthouse, you'll run out of beach; passage is blocked by waves crashing against a rocky point. However, 100 yards before the point, a cut in the cliffs will lead you to a crumbling dirt road to the bluffs above the ocean. Once on top, you can follow a number of cow trails toward the lighthouse, established by the federal government in 1855 and still in use.

Stay away from the lighthouse and Coast Guard Reservation; visitors are not welcome. But enjoy the terrific view.

Return the same way. If you want, descend the bluffs just north of the lighthouse to one of the handsome isolated coves and work on your tan.

Tiger Lily

Point Sal Trail (SB-8)

Point Sal State Beach to Point Sal; 5 miles RT
Point Sal State Beach to Rancho Guadalupe Dunes Co Park; 12 miles RT

When your eye travels down a map of the Central California coast, you pause on old and familiar friends—the state beaches at San Simeon, Morro Bay and Pismo Beach. Usually overlooked is another state beach—remote Point Sal, a nub of land north of Vandenburg Air Force Base and south of the Guadalupe Dunes. Windy Point Sal is a wall of bluffs rising 50 to 100 feet above the rocky shore. The water is crystal clear and the bluffs provide a fine spot to watch for whales.

Point Sal was named by Vancouver in 1792 for Hermenegildo Sal, at that time commandante at San Francisco, in recognition of favors received by him. The state purchased the land in the 1940s. Whatever administration there is, is accomplished by Santa Barbara County. There are no facilities whatsoever at the beach, so remember, if you pack it in, pack it out.

This hike travels Point Sal Beach, then takes to the bluffs above rocky reefs. At low tide you can pass around or over the reefs; at high tide the only passage is along the bluff trail. Both marine life and land life can be observed from the bluff trail. You'll pass a seal haul-out, tidepools, sight gulls, cormorants and pelicans, and perhaps see deer, bobcat and coyote on the open slopes.

The hike follows a narrow bluff trail and should not be undertaken by novice hikers or those afraid of heights.

Directions to trailhead: Turn west off Highway 12, three miles south of Guadalupe on Brown Road. In a few miles Brown Road ends at a ranch. Bear right on Point Sal Road, partly paved, partly dirt washboard (impassable in wet weather). Follow this road 5 miles to its end at the parking area above Pt. Sal Beach.

The Hike: From the parking area follow the wide trail through chaparral down the bluffs to the beautiful crescent-shaped beach. Hike up-coast along the wind-swept beach, which usually hosts a few families and their tents. In ¼ mile you'll reach the end of the beach at a rocky reef, difficult to negotiate at high tide. A second reef, encountered shortly after the first, is equally difficult. Atop this reef there's a rope secured to an iron stake, to aid your descent to the beach below. The rope is particulary helpful in ascending the reef on your return.

Unless it's very low tide you'll want to begin following the narrow bluff trail above the reefs. The trail arcs westward with the coast, passing through wild strawberry and giant coreopsis and occasionally dips down to rocky and romantic pocket beaches sequestered between reefs. Fishermen cast from these beaches and high rocks and report good fishing.

About two miles from the trailhead, you'll hear the bark of seals and get an aviator's view of Lion Rock. Harbor seals, often called leopard seals because of their silver, white, or even yellow spots, bask on their large offshore rock.

Continuing on the bluff trail past sand verbena and cow lilies, the trail dips down near a seal haul-out. (Please don't approach or disturb these creatures.) You'll then ascend rocky Point Sal. From the point, you'll view the Guadalupe Dunes complex to the north and the sandy beaches of Vandenburg Air Force Base to the south. The trail passes behind the point, joins a sandy road, and descends to a splendid sandy beach north of the point.

Our trail is almost due north now on a nearly two-mile-long sandy beach. The beach is almost always deserted except for a few fishermen and lots of pelicans.

At the north end of this unnamed beach is a rocky reef, difficult to climb. A hundred yards south of the reef you can pick up a bluff trail which passes above the reef and looks down on a tiny waterfall that trickles down to the beach. Continue on this breaktaking, but sometimes precarious bluff trail. You'll look down on red-tailed hawks riding the updrafts and watch the ocean boil up over the reefs.

The rocky bluffs and trail come to an end as you approach the sand dunes south of Guadalupe Dunes County Park. You can walk the sandy beach to the county park near the mouth of the Santa Maria River. Here you may join the CCT and hike to Oregon or return the same way to Point Sal Beach.

San Luis Obispo County

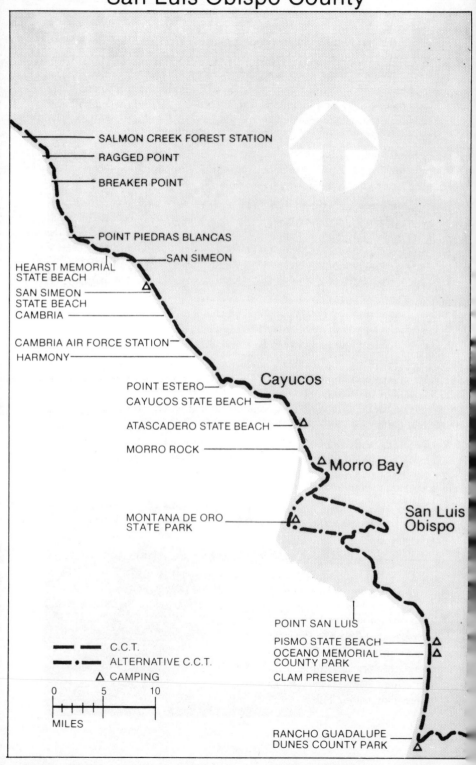

San Luis Obispo County

CCT hikers can travel 100 miles along the San Luis Obispo County Coast, hiking across wide sandy beaches, around protected bays, and over rugged headlands. Hikers may see birdwatchers gazing at the shimmering tops of eucalyptus trees where great blue herons build their nests, rockhounds gathering moonstones, clammers digging for Pismo and Razorback clams. The county's magnificent coastline offers all these natural attractions and more.

The southern part of the county's coastline is dominated by wind-swept sand dunes, great heaps of sand held in place by ice plant, verbena, grasses, sea rocket and silver lupine. The longshore currents that normally carry sand along the coast is interrupted by rocky headlands and the sand stays in the local area, later to be deposited by the wind on the dunes above the beach.

Morro Bay and its adjacent mudflats is an amazingly fertile wetland, one of the largest and most significant wildlife habitats on the California coast. The bay ranks within the top ten areas in the nation in terms of numbers of bird species sighted in a single day. Guarding the bay is the much photographed Morro Rock, "The Gibraltar of the Pacific," which stands halfway between Los Angeles and San Francisco.

North of Morro Bay tall bluffs rise above the beach. Land and sea blend into one astounding tableau. Cliff trails follow the edges of grassy coastal terraces, past Cambria Pines, San Simeon, Piedras Blancas, to just south of the Monterey County line. Unlike the theatrical cliffs of Big Sur, these bluffs are accessible; they ebb and flow toward the Coast Highway like the tide, sometimes 20 feet away, sometimes a mile, and always an adventure for those hiking along Land's End.

CCT at a glance

TERRAIN: Sandy beach and dunes in south coast, rolling pastureland around Montana De Oro State Park. North of Cayucos bluffs rise above the water. Trails travel over grass-covered coastal shelf in north county.

OBSTRUCTIONS: Diablo Canyon Nuclear Power Plant locks up ten miles of pristine coastline and requires an inland detour. Private property along Point Estero can present problems. Beyond Ragged Point to the county line, sheer cliffs mean a detour along Highway 1.

TRANSPORTATION: Amtrak stops in San Luis Obispo. For bus transit information, call (area code 805): Coastlines Bus Service, 772-3562; San Luis Obispo County Area Transit (SLOCAT) 549-5252; South County Area Transit (SCAT), 489-5400; and North Coastal Transit, 544-2500. Buses for all systems can be boarded at regular stops or flagged at any corner or safe location.

CAMPGROUNDS/ACCOMMODATIONS:
San Simeon State Beach
Atascadero State Beach
Morro Bay State Park
Montana De Oro State Park
Pismo State Beach
Oceano County Park
Oso Flaco Lake

CCT: SAN LUIS OBISPO COUNTY

CCT enters San Luis Obispo County just north of the mouth of the Santa Maria River, a wetland area where several endangered species reside, including the California least tern and brown pelican. Back across the river in Santa Barbara County is Rancho Guadalupe County Park, where you'll find the highest sand dune on the west coast, 450-foot tall Mussel Rock.

Brightening the dunes on your way north is yellow and magenta sand verbena, coreopsis, daisies and white-fringed aster. Flowers, plants and grasses are vital to the dune ecosystem. Some dunes continue to be formed today. The active, moving ones are those with little or no vegetation. As you near Oso Flaco Lake, where Portola camped on his expedition, your quiet thoughts may be interrupted by a dune buggy.

Stretching from Oso Flaco Lake north to Pismo Beach are the Oceano Dunes, the highest and whitest sand dune formations in California. These dunes evolved many thousands of years ago, between Ice Ages, through deposition by the Santa Maria River, and the sculpting of wind and sea. The cliffs of Point Sal acted as a sand trap to keep the dunes from straying south.

A mile inland from CCT, nestled in the dunes, are several shallow lagoons and marshes, including Black, White, Willow and Pipeline Lakes. Too shallow for fish, they provide nesting sites for mallard and teal.

During the Great Depression, the dunes were home to the "Dunites," a motley collection of writers, artists, hermits, nudists and astrologers who lived in driftwood shacks and published their own magazine called *The Dune Forum*. The dunes were a featured location in the 1964 movie, "The Great Race."

Weekdays are apt to be calm among the dunes, but weekends bring batallions of off-road vehicles. Vehicles are permitted on the beach from Oso Flaco Lake north a few miles to Pismo Dunes Preserve. At the Preserve, hikers have the undeveloped dunes to themselves, sans ORV's. Beyond the Preserve is the little town of Oceano, one of the "Five Cities" of southern San Luis Obispo County. In 1904, Oceano boasted beach cottages, a wharf, and mammoth La Grande Beach Pavillion. The developer's grandiose plans of turning Oceano into a tourist mecca did not materialize and pavillion, wharf, and cottages were buried beneath advancing dunes.

Oceano Memorial Park offers campsites near Oceano Lagoon. Adjoining the lagoon on the east is Pismo State Beach's Oceano Campground. A mile north CCT passes a second campground operated by Pismo State Beach—North Beach Campground, located 300 yards inland on the other side of the sand dunes.

Pismo clam

Common littleneck clam

Pismo Beach, "The Clam Capital of the World," has been famous for its dunes and succulent clams since before the turn of the century. Every summer, from 1895 through the 1920s, a tent city sprang up, populated by inland dwellers fleeing the heat. They paid $8 a week to rent one of the green-and-white-striped tents clustered around a dance pavillion.

ON PISMO HILLS

By Katherine Lynch Smith

The ocean stretches before me,
    The green hills lie behind,
And the scent of the flowers comes o'er me,
    Borne on the breath of the wind.
Oh, Hills, in your vernal beauty!
    Oh, Ocean, so calm and so still!
Lead this restless heart to duty,
    And strengthen this wavering will.
Let me learn from your might and your power
    To be, too, steadfast and strong;
Ne'er swayed by the change of the hour,
    Ne'er turning from right to the wrong.

The Pismo Beach Chamber of Commerce won't say there are no more Pismo clams, but even the most optimistic beach booster admits the catch has been drastically reduced. Sea otters are responsible for some of the decline. Otters float on their backs and use rocks to break open the clams. A single otter can consume 20-25 pounds of shellfish a day. Human clam diggers, who've been visiting Pismo for decades, have taken a greater toll.

CCT continues past the 950-foot long Pismo Beach Pier to Shell Beach, once the site of Chumash Indian shell gathering. Skin divers explore the waters and rock coves of the area. A community of harbor seals bask on the rocks and sea otters patrol kelp beds off shore.

Depending on the tide, CCT hikers will inevitably be forced off the beach by natural and man-made obstacles in the vicinity of Pismo Beach's "Motel Row." You'll follow business Route 1 (Shell Beach Road) through town. Several residential streets allow beach access, but a Bavarian castle and a few other beach houses make through passage extremely difficult. It's easier continuing on Shell Beach Road, passing the residential community of Sunset Palisades. Just beyond the junction with El Portal Drive, you'll see a sign on your left that says *Welcome to Pismo Beach* and a road with a gate across it bearing another sign, *Closed Road*. This closed road, Cave Landing Road, is really a bridle path and CCT follows it behind a row of houses for a few hundred yards west. The road, reaches the bluffs overlooking the ocean and turns north. Cave Landing Road, now paved and open to vehicle traffic, continues along the bluffs for a mile to Avila Road and Avila Beach.

(If you're ready to camp for the night or wish to avoid Avila Beach, continue on Shell Beach Road, which shortly turns into Palisades Road, then make a left on Avila Road. Here you'll find Avila Hot Spring Campground, mostly geared for the RV crowd, but tenters are welcome. The hot springs here and those ¾ mile farther down the road at Sycamore Mineral Springs, are a delight for footsore, bone-weary hikers.)

Following Cave Landing Road, CCT stays atop the bluffs and soon overlooks Pirate's Cove Beach. There's access from the handsome rocky cliffs on the north. Union Oil tanks and barbed wire prevent passage over Fossil Point directly into Avila Beach, so continue following the road another few hundred yards past Pirate's Cove to Avila Road. Turn left. On the right of the road is San Luis Creek, which passes through a private golf course. Black-crowned night herons roost along the creek. In a half mile you'll reach Avila Beach.

Avila claims the warmest beach climate along the central coast because it's protected by a ring of mountains. The name Avila honors a former corporal of Mission San Luis Obispo. The business district seems a charming relic from the late 1940s.

Beyond the little town is the Union Oil Pier, Port San Luis Beach and the quarter-mile-long Port San Luis Pier, damaged severely in the winter of '83. Pier fishermen try for red snapper, perch, blue gill and rock crab.

North of Avila Beach, the ten miles of coastline between Port San Luis and Point Buchon is the domain of sea lions, cormorants and pelicans; it's one of the most beautiful and pristine sections of the California coast. Unfortunately, it's completely inaccessible due to the construction of the Diablo Canyon Nuclear Power plant, built atop an earthquake fault six miles north. Until the dispute over the controversial plant is settled, hikers are forced to detour away from the coast to reach Montana De Oro State Park.

We leave Avila Beach on Avila Road, heading inland. CCT bears left onto San Luis Bay Drive and then shortly makes another left to the northwest on Sea Canyon Road. In the canyon are orchards, small farms and ranches. In season, fruit is sold at roadside stands. In the springtime, the road is lined with the purple of lupine and giant thistle and the orange of California poppies. After four miles, the pavement ends and joins Perfumo Road. In a short distance a rough side road is offered on the left. It's possible to hike to Montana De Oro State Park from this turnoff via a series of power line roads, mining road, cow trails and ranch roads. However, this route crosses private property and an easement has not yet been secured by the California Coastal Trails Foundation; therefore it's not described in this guide. In addition, the California Coastal Trail when completed, will be following a more inland parallel course through the Santa Lucia Mountains, thus in a few years making this backdoor route to Montana De Oro of interest only to day hikers. For an update on the situation, contact the California Coastal Trails Foundation.

Those skilled hikers wishing to hike from Perfumo Road to Montana De Oro State Park, who arm themselves with the appropriate topos, should be able to improve a route to the eastern boundary of the state park on Islay Creek. Hint: Stay out of the Coon Creek drainage. Second hint: The hilltops crossed by electrical transmission lines provide good views and simplify the orienteering process. If you decide to cross private property, be on your best behavior. No camping, no fires. Don't bother the grazing livestock.

CCT hikers not wishing to cross private property, will continue a few more miles on Perfumo Canyon Road, turning left on Clark Canyon Road and following this road to Los Osos Valley Road. A left turn takes you westward through the town of Los Osos and on to the beach on Estero Bay. The hiker may then double back south along the beach and bluffs to Montana De Oro State Park or head north onto Morro Bay sandspit.

This guide resumes the trail description just outside the eastern boundary of Montana De Oro State Park on Islay Creek. No camping or fires and please respect private property.

CCT descends with the road which hugs Islay Creek as it runs from the San Luis Mountains to the sea. CCT follows the road, first through the ranch, then through Montana De Oro State Park. Yellow tree poppies splash the pastureland from the ridgecrest down. The road passes through the San Luis Range, the fist of mountains extending between Avila Beach and Morro Bay. The San Luis Mountains encompass Montana De Oro State Park, Irish Hills and Diablo Canyon. Several high points provide bird's eye views of the central coast: Green Peak, Valencia Peak, Saddle Peak, Bald Knob, Mine Hill. These peaks top out at less than 2,000 feet.

Three miles from the Pacific, CCT reaches Montana De Oro State Park boundary. Climb over the locked gate and continue down the well-maintained dirt road. You'll soon pass an abandoned barn. Along the road are huge coffeeberry plants, purple nightshade and flowering sage.

You may get a whiff of sulphur from the hot springs concealed in the impenetrable undergrowth lining the creek. 1¾ miles from the park boundary, look for an unsigned side trail on your left, which descends a steep ravine to two Islay Creek waterfalls.

CCT continues its gradual descent along the road. As it nears the ocean, it is lined with thistle, mustard and wild cucumber. Across Islay Creek, the park campground can be seen.

When you reach the paved park road, bear left a few hundred yards to the park entrance. Across from park headquarters (the former Spooner Ranch home built in 1892) is Spooner Cove, once a landing place for bootleggers. Atop the bluffs bookending the cove grow fields of mustard and poppies which give the park its "Mountain of Gold" name.

The 50-space park campground stretches along the bluffs above Islay Creek. There's a large grassy group camp area.

Before leaving Montana De Oro State Park, CCT hikers planning to follow the trail northward onto Morro Bay Sandspit should check in with the U.S. Coast Guard (772-3202) or Morro Bay Harbor Office (772-1214) and inquire about the on again-off again Clam Taxi Service, a small boat that will take you from the end of the sandspit a short distance across the estuary to the mainland. A fee is charged and service is sporadic. The only pay phone enroute is at park headquarters.

CCT climbs the bluffs just north of Spooner's Cove, dipping in and out of ravines on a series of horse and surfer trails. Your improvised route provides fine overlooks of handsome (and accessible) pocket beaches and rocky coves. If it's low tide, you may wish to hike for stretches along the beach.

Blufftop trails bring you to narrow Hazzard Canyon, where you join an unsigned trail, following it through aromatic eucalyptus, blessed in season with congregations of monarch butterflies, down to the beach. Along the shore of Hazzard Canyon Beach are tidepools, good swimming and superb surfing. Also at the beach are thousands of wave-polished sandstone rocks with swiss-cheese-like holes in them. These holes are bored by the piddock, an industrious member of the mollusk family.

CCT follows either bluff or beach to the southernmost tip of Morro Bay. We head over low dunes to the ocean side of the sandspit. The spit varies in width and is a little less than ½ mile wide at its widest part. Atop some of the higher dunes (about 80 feet above sea level), you'll be treated to good vistas of the bay, Morro Rock, and nearby mountains.

Approximately 1½ miles from the north end of the spit, in an area of little vegetation, climb over the dunes to the east toward Morro Bay. Here you'll find sheltered Houseboat Cove. Across the bay you'll spot the Morro Bay State Park Museum.

Sanderling

Hike along the bay shore. Many birds such as sandpipers and curlews feed in the mudflats. The vegetation increases, with saltgrass and buckwheat predominating. Continue hiking northwest along the shore until you reach the north end of the spit. Here, one hopes, you'll catch the Clam Taxi over to the town of Morro Bay.

From the landing CCT hikers improvise a route past the fish markets, open air boiling pots and fish-'n-chip stands along the embarcadero. Dominating the seascape of Morro Bay is the "Gibraltar of the Pacific," 576-foot high Morro Rock, first sighted by Juan Cabrillo in 1542. It's quite a challenge to frame a picture of the rock without including the unsightly, sky-scraping smoke stacks of the Pacific Gas and Electric Company's power plant at water's edge.

Atop Morro Rock roosts the endangered peregrine falcon, the quickest and most prized of falcons. The falcons are staging a comeback from the devastating effects of DDT which caused them to lay thin, fragile eggs. The 50 million-year-old volcanic peak was used as a rock quarry as late as 1969, but is now a wildlife preserve and part of the state park system.

CCT angles around the northwest side of the rock, fords the usually shallow mouth of Morro Creek and begins crossing wide and firm Atascadero ("Mudflat" in Spanish) Beach. The state park campground, about a mile upbeach from Morro Rock, isn't very accommodating to hikers, for it is primarily a blacktop parking lot of RV campers, each having its own 20 by 30 foot space. However, it should be noted this is the last official campground until CCT reaches San Simeon State Beach, 25 miles north.

Leaving Atascadero State Beach, CCT travels over a brief stretch of rocky coast having some fine tidepools. Highway 1 comes close to the beach for a short distance but soon veers off to a respectful distance. North of the rocks it becomes easy walking on hard-packed sand. This stretch of beach is heavily used only once a year when competitors in the Morro Rock to Cayucos Fun Run sprint by. Often otters and harbor seals can be seen just beyond the surf line. About halfway along this route you'll pass the private Standard Oil Pier and begin traveling over the sands of Morro Strand State Beach. A nearly continuous (but not obstructive) line of houses extends north from here along Cayucos Beach and Cayucos State Beach to the north.

At the unincorporated community of Cayucos, CCT approaches a fishing pier. The little town is the last place on the coast to purchase supplies en route north to San Simeon. After the breakup of the Mexican land grant Morro y Cayucos Rancho, ranchers and dairy farmers began working the land. A wharf was built to ship their products and Cayucos came into being about 1875 as a seaport.

Standing on the pier, you may see sea otters frequenting the waters around the pilings. As you look up-coast you'll see the rocky sea cliffs that rise at the north end of town and continue all the way to Monterey. While contemplating the cliffs, you must decide whether you want to hike the bluffs, and cross some private property while viewing one of the most spectacular and pristine sections of coastline along the CCT. The alternative is to hike along Coast Highway, which is legal but in many cases dangerous because of the lack of shoulder. The bluff route is described below.

CCT leaves Cayucos on North Ocean Avenue, then resumes on the bluffs. You may be sharing the dramatic coastline views with bovine friends. If you proceed along-side Highway 1, there's a ranch road paralleling the highway for a while. You'll pass a dirty campsite with a fire ring. Emblazoned in charcoal on the rocks above is a message: *Pick Up After Yourself.*

Rounding Cayucos Point, you'll pass an old abandoned trailer. Continue walking along the bluff edge. As you approach the next prominent point, Point Estero, you'll ford the mouth of Ellysly Creek, passing over a beautiful driftwood beach.

The next few miles are difficult. From here, you ascend the bluffs on a ranch road, improvising a path on cow trails. The next few miles will take you as far away from the highway as you'll get in northern San Luis Obispo County. The coast here could be termed "The Lost Coast"; certainly it's lost to motorists whizzing by on Highway 1 and lost to the public since there's no official vertical access for over 15 miles. As you improvise your route, keep in mind that the power lines you occasionally follow do not always lead north-south and are not to be relied upon for directional puropses. Be prepared to dip in and out of or head inland around several steep ravines.

In a few miles, you'll spot atop high bluffs the now decommissioned Cambria Air Force Station. The station, a radar facility with lots of barracks, is now being considered by the American Youth Hostels Association as a possible hostel. Coastline views from the base are superb.

CCT hikers must negotiate a high fence at the northern perimeter of the base. A powerline road follows the ridge crest northward toward Cambria. On the east side of the road, behind a fence, are cows and prize bulls. To the west is unfenced rangeland on table top bluffs overlooking the ocean. The rough power line road enters the outskirts of Cambria Pines. These are the first pines we've seen along the coast since the rare Torrey Pine in San Diego County. As you enter the pines, the most expeditious route back to the coast is to hike cross-country northwesterly through the pines toward the ocean.

You'll end up at the residential area of Cambria Pines Manor, then follow Richard Avenue to Ardath Avenue, turning west to Drake Street, then turning right and hiking a few blocks to Sherwood Drive and coastal access.

At the north end of Sherwood Drive, an excellent bluff trail begins, heading north through pastureland toward a second Cambria Pines neighborhood. You'll pick up Windsor Boulevard for a short distance, taking the first through-street west to Nottingham and continuing along Nottingham. There's a coastal access here off Nottingham from undeveloped state park land, and sporadic bluff trail. The trail descends to the beach at Shamel County Park.

From the county park, walk north along Moonstone Beach, which at this point is composed of colored pebbles instead of sand. In a quarter mile, you'll ford Santa Rosa Creek. The creek mouth is sometimes difficult to cross at high tide and after winter storms. The beach narrows. A little less than a mile from the trailhead, climb the low bluffs and begin following a series of unmarked trails through coastal sage atop the bluffs. You may see sea lions sunning themselves at a rocky cove below.

When you reach the old highway bridge, cross the usually dry creek. Just above the creek is the picnic area at Leffingwell Landing, nestled in a sheltered cypress grove. During the 1870s and 1880s, ships unloaded lumber and other goods here for pioneer settlers along San Simeon Creek.

CCT picks up again at the bluff edge past the picnic area. Soon you'll be treated to views of Piedras Blancas Lighthouse to the north and Hearst Castle inland. Indian mortar holes are ground into the tops of the sandstone bluffs rising above the beach. The trail passes behind a grove of cypress trees and soon descends to the beach.

You may follow the beach or a bluff trail for the next mile north to San Simeon State Beach. As you approach the campground, turn east under the highway bridge to the special hiker-biker campground on your left. For some reason, park rangers crowd all the hikers and bikers into two campsites near the highway. However, it's the only official campground from Morro Bay to the Los Padres National Forest and beggars can't be choosers....

Leaving San Simeon State Beach, walk under the highway bridge back to the beach. Keep to the north side of San Simeon Creek to avoid fording it. The beach soon ends and you slip and slide for fifty yards over some kelp-covered rocks. Pass a pink house, ascend the bluffs and begin following them (no trail, but easy walking). Stay with the cliff edge through its meanderings as CCT roughly parallels the highway, about ¼ mile away. 1½ miles of bluff hiking brings you to a private driveway and the outskirts of San Simeon. The town is a strip of gas stations, motels and restaurants fronting both sides of Highway 1.

Leaving town, there's not much bluff trail. You must bushwhack or follow the highway shoulder until you reach the first blue and white "Viewpoint" sign. From the turnout, descend to the beach. At all but highest tide, it's possible to hike William Randolph Hearst Beach to Hearst Pier at San Simeon Bay.

San Simeon Bay provides fairly good refuge from northwest and west winds. Fishing boats work the water near here for albacore in late summer and fall. In 1951 after the death of William Randolph Hearst, his estate donated land south of the general store for a park. In 1957 San Luis Obispo County built the fishing pier. Later, the state took over operations of the park.

The land around San Simeon was once a rancho belonging to Mission San Miguel. Leopold Frankl, a German-Austrian emigrant opened the general store in 1874 and founded and named the town. He sold his land to Senator George Hearst, who then began developing the family estate.

CCT passes a picnic ground located in the eucalyptus grove just north of the pier. Follow the beach until it begins to arc westward, then ascend the bluffs to a narrow dirt road leading atop the wooded bluff. This road, narrowing to a trail, provides coastline and castle views, passing a dilapidated shack and curving toward San Simeon Point. At the Point are additional breathtaking views to the south of the undeveloped San Luis Obispo County coast. CCT continues around the point on overgrown blufftop trails, then passes under the boughs of cypress on a dark tunnel-like trail for ¼ mile, before re-emerging back on the bluffs. The bluff trails grow more faint and erratic as you descend a low sand dune to the beach.

Follow the beach until tides and rocks prevent forward progress and you meet the coast highway. Continue for another two miles along the shoulder of the highway, following bluff trails from one turnout to another, from five feet to two hundred feet from the highway. Near where you sight a triangle-shaped modern sculpture you'll head inland, leaving the highway behind. A

mile north of the sculpture CCT crosses a chalky peninsula named Piedras Blancas, or white stones. A 74-foot high brick tower, Piedras Blancas Lighthouse casts its powerful beam 20 miles out to sea to warn mariners of this treacherous rocky coast. Piedras Blancas has been in continuous operation since 1874 and is now fully automatic. The Coast Guard operates the lighthouse, not open to the public. As you'll soon see, the lighthouse is one of the few exceptions to Hearst ownership along this section of coast.

From Point Piedras Blancas to a bit south of Ragged Point stretch ten miles of blufftop trail. The trail rarely travels more than ½ mile from the highway, but you'll feel a surprising isolation as you hike the trackless beaches and majestic bluffs. From Piedras Blancas CCT follows bluff trails for three miles, then descends to a magnificent beach at the mouth of Arroyo de la Cruz. Ford the stream mouth here and depending on the tide, sprint around or climb over rocky Point Sierra Nevada. With a little imagination you can guess how the point got its name: The ocean here crashes into what looks like the towering east walls of the High Sierra.

CCT follows the beach on the northern side of the point. Delay ascending the bluffs again for a few hundred yards because poison oak flourishes up there. A little south of where the bluffs change from white to dun-colored, clamber up the bluffs.

Atop the bluffs are superb, well-worn cow trails. Hike back near the highway for a brief stretch, then leave it as you pass north of steep Arroyo de los Chinos. Occasionally cattle graze near this stretch of trail.

Bluff trails grow more faint as you go north and vegetation increases. You'll do some bushwhacking. CCT rounds Breaker Point which points at some handsome offshore castellated rocks. You then negotiate steep Arroyo Hondo Ravine. Beyond the ravine, look sharply for an east-west feeder trail. If you take the trail east you'll arrive back on Highway 1 about 2½ miles south of Ragged Point. If you go east, you'll soon arrive at a place where the trail descends the bluffs to the beach. CCT follows this seldom-walked beach to the mouth of San Carpoforo Creek. If it's low tide (high tide— walk the highway), pick your way over the rocks for 1½ miles to a nature trail which leads up the cliffs to Ragged Point Inn. A burger stand, motel and phone greet you here.

CCT follows the highway, climbing steeply to the north and leaves the shoreline far below. In 1½ miles CCT crosses into Monterey County, in 3 miles enters Los Padres National Forest, and in 4 miles reaches the Salmon Creek Forest Station and the Salmon Creek trailhead. Here CCT turns inland into the coast range.

Because Highway 1 from Ragged Point to Salmon Creek has little shoulder, you might consider taking the Coastlines Bus or trying to catch a ride north.

Nipomo Dunes Trail (SLO-1)

Oso Flaco Lake to Santa Maria River; 8 miles RT

The Nipomo Dunes are one of the largest relatively undisturbed dune complexes in California. The Nipomo-Pismo Dunes run from Pt. Sal just north of Vandenburg Air Force Base to the northern end of Pismo State Beach in San Luis Obispo County. In 1974, this 18-mile stretch of oceanfront in Santa Barbara and San Luis Obispo Counties was declared a national landmark by then Secretary of the Interior Rogers Morton. The Federal Registry of Natural Landmarks registers sites that provide significant illustration of our nation's natural history. Areas must have been maintained in relatively undisturbed condition. Landmark status has not
This trail departs from Oso Flaco Lake, a 75-acre lake and marshland, and travels along the dunes to the mouth of the Santa Maria River.

Directions to trailhead: From Highway 1, 3 miles north of State Highway 166, turn west on Oso Flaco Road, driving 3.6 miles to road's end at the dunes.

The Hike: Largest of the lakes that dot the dunes, Oso Flaco offers fishing for catfish, crappie and sunfish. The Portola Expedition of 1760 camped at Oso Flaco on September 2-3. The soldiers killed a huge bear and feasted on it. Although Father Crespi, diarist and spiritual counselor for the expedition wanted to call the lake, "Lake of the Martyrs, San Juan de Perucia and San Pedro de Sacro Terrato," the soldiers' more humble name of Oso Flaco or "lean bear" stuck.

From the lake you'll walk oceanward up and down the dunes to the water's edge and head south. Occasionally, your quiet thoughts may be interrupted by a stray dune buggy.

Brightening the dunes is yellow and magenta sand verbena, coreopsis, daisies, and white-fringed asters. The flowers and grasses are vital to the stability of the dune ecosystem. Some dunes continue to be formed today. The active, moving ones are those with little or no vegetation.

Four miles from the trailhead you'll reach the mouth of the Santa Maria River, a wetland area. Residing near the river mouth are several endangered birds including the California least tern and the California brown pelican. An endangered plant, the surf thistle grows here.

Across the river is Rancho Guadalupe County Park, where you'll find the highest sand dune on the west coast, 450-foot tall Mussel Rock.

Return the same way.

Coon Creek Trail (SLO-2)

Pecho Road to Old Shack; 5 miles RT; 200′ gain

Coon Creek is a year-round creek that winds through the Irish Hills along a lush canyon to the sea. The vegetation is so thick in the canyon that hikers often pass within a few feet of the creek, hear its murmuring, yet are unable to see it. Ancient bishop pines line the banks of Coon Creek. The canyon teems with wildlife—black-tailed deer, rabbits, possum, and, of course, raccoons.

This hike follows the creek, crossing it a half dozen times. Occasionally, the trail passes through meadowland full of fiddle-neck, poppies, mustard, and monkey flowers. Beware: An extraordinary amount of poison oak grows along this trail.

Directions to trailhead: From Highway 101, exit on Los Osos Road, continuing northwest for 12 miles until the road turns south to become Pecho Road. Pecho Road leads to the park. Continue four miles past the park entrance sign to road's end. A parking area is located at the trailhead.

The Hike: From the trailhead, you descend for a moment into a shallow canyon, then climb a ridge for a brief shoreline glimpse. The trail soon ventures into Coon Creek Canyon. You can hear but not see the creek on your right. The trail is choked with maple, willow, mugwort, poison oak. A half miles from the trailhead stand some Bishop pines.

Crossing and re-crossing the creek, you pass live oaks covered with moss. At the two-mile mark, you ascend a short way into an exposed grassland that can either be considered a curse on hot days or a welcome response from the poison oak. The meadow displays abundant wildflowers in spring. On your left you'll spot the faint, unsigned Oats Peak Connector Trail. Trail's end occurs in a mixed stand of old oaks and cedars. Here you'll find the crumbling remains of an old shack.

On old topos, the Coon Creek Trail is shown extending for several more miles along the creek. Alas, you can continue but another .1 mile before hiking headlong into impenetrable thickets of poison oak.

Return the same way.

Montana De Oro Bluffs Trail (SLO-3)

Spooner's Cove to Grotto Rock; 3½ miles RT

At the turn of the century, the greater portion of what is now known as Montana De Oro was part of the Spooner Ranch. The most popular beach in the park is Spooner's Cove, whose isolation made it an ideal landing spot for *contrabandistas* during the Mission era and for bootleggers during Prohibition.

The trail debarks from the south side of Spooner's Cove and travels atop rugged cliffs. The park service has placed several signs which state the obvious: *"Danger, Sheer Cliffs."* Atop the bluffs grow fields of mustard and poppies, which give the park its "mountain of gold" name. Side trails drop to Spooner's and Corallina Coves, with fine tidepools to explore.

While hiking the bluffs, you may see harbor seals venturing ashore or otters diving for food beyond the surf line. Bird watchers delight at the pelicans, albatross, cormorants and the crimson-billed black oyster catchers.

Directions to trailhead: From Highway 101, exit on Los Osos Road, driving 12 miles northwest. The road turns south to become Pecho Road and leads into the park. Drive 2.7 miles past the park entrance sign, a hundred yards south of park headquarters, and park on the west side of the road at the unsigned trailhead.

The Hike: The trail crosses a dry creek on a footbridge. You'll soon pass a sign telling the history of Spooner's Cove, named after William Bradford Spooner, who settled here in 1892. Perched on the bluff above the cove is the foundation of Spooner's warehouse. Spooner loaded his ranch products by means of a chute onto waiting ships below. As you approach the western edge of the cove, Morro Rock pops into view on your right. Below, seals are often seen.

A half mile from the trailhead, a fork to the right leads to Corallina Cove, bedecked with sea-polished broken shells and beautiful beach pebbles. Scuba divers enter the cold, clear waters here. The crystal-clear tidepools are full of anemones, starfish, mussels, and colorful snails.

Returning to the Bluffs Trail and staying to the right, you'll cross a wooden bridge. A mile from the trailhead is Quarry Cove, also with fine tidepools. A half mile further along the bluffs, covered with splotches of purple lupine and giant thistle as well as the golden flowers, brings you to an overlook above some sea caves. Beyond is Grotto Rock.

You may return the same way, or bear left and return via Pecho Road.

Valencia Peak Trail (SLO-4)

Montana De Oro State Park Headquarters to Valencia Peak
4 miles RT; 1300' gain

From a distance, you might suspect that 1345' Valencia Peak is one of the morros — those distinct cone-shaped mountains that dot the San Luis Obispo County Coast south of Morro Bay. However, Valencia Peak rose out of the sea in relatively recent geologic time. Geologists believed the processes that fashioned the mountain—tilting, folding, and upheaval occurred only five million years ago.

The peak's oceanic origins are revealed by its upper slopes, which were once beaches. You can find strands of beach sand and rocks that have been bored out by clams. Atop the mountain are fossil shells.

This hike switchbacks over what were once sea cliffs to the top of Valencia Peak, named after an Indian family who lived nearby in the year after the Mission period. On fog-free days, the view of the Central coast from Point Sal to Piedras Blancas is inspiring.

Directions to trailhead: See Montana De Oro Bluffs Trail for directions to Montana De Oro State Park. Valencia Peak Trail begins at the signed trailhead across the campground road from park headquarters.

The Hike: The trail follows a (usually) mowed stretch of grass along Pecho Road to the south a few hundred yards, then turns inland and starts upslope to the east. Already, you can distinguish the series of marine terraces on the mountain. In Spring, lupine, mustard, Indian paintbrush and a host of wildflowers cover the coastal slope.

A half mile long, the trail bears left and heads directly toward the peak. You dip in and out of a dry gully and begin switchbacking over outcroppings of Monterey shale, traces of former sea cliffs.

As the trail levels out you'll go left at a trail junction, then begin switchbacking again, more steeply this time. The trail forks a bit below the peak; both trails go to the summit. Here you'll look out over the (mostly) unspoiled central coast. You can see the twenty-million-year-old volcanic peaks of Morro Rock, Hollister Peak, Black Mountain.

Look for fossils, enjoy the view, and return the same way.

Morro Bay Sandspit Trail (SLO-5)

Woodland Avenue to North end; 5 miles one way

Scientists estimate 80% of all sea life along the central coast originates in Morro Bay Estuary. The phenomenally nutritive waters of different salinities mix, creating an amazingly fecund environment. The triangular-shaped marsh, lined with eel grass and pickleweed, is an important spawning and nursery habitat for such fish as the California halibut and sand perch. Beneath the surface of the bay are oysters, clams, worms, snails, crabs and shrimp.

This trail follows the sand dunes and ridge that separate Morro Bay on the inland side and Estero Bay on the ocean side. Mudflats on the bay side of the spit are covered at high tide and mucky at other times, so you'll probably want to hike along the ocean side, as suggested in this trail description.

The spit varies in width and is a little less than a ½ mile wide at its widest. Atop some of the higher dunes (about 80 feet above sea level), you'll be treated to good vistas of the bay, Morro Rock, and nearby mountains.

Directions to trailhead: From Highway 101 in San Luis Obispo, exit on Los Osos Valley Road, traveling to its end. A block after the road curves left and becomes Pecho Road, turn right onto Woodland Avenue. Drive to road's end and park. A key element to this hike can be the Clam Taxi, a water-taxi service between the town of Morro Bay and the north end of the sandspit. Service is sporadic and you would be advised to call the U.S. Coast Guard (772-3202) or the Morro Bay Harbor Office (772-1214). The Clam Taxi was at the foot of Pacific Avenue, right off of Embarcadero. If you can arrange a car shuttle and utilize the Clam Taxi, you can make this a one-way hike or begin from Morro Bay. A canoe rental service is also available on the mainland. It might be best to canoe over to the spit and hike north-south.

The Hike: Walk through the eucalyptus grove on a dirt road, bearing left at the first fork, then right at the second. When you reach the dunes ascend on a steep diagonal. From the top of the dunes are fine views of the bays and Shark's Inlet.

From the crest of the dunes, descend into a valley. You'll see a large shell mound in the center of the valley, a massive artifact left by the Chumash Indians. They piled clams, cockles, snails, and even land game in these kitchen middens. Inspect this shell mound and the others on the spit with care. The bountiful marsh is so full of bird, land, and aquatic life, it's easy to imagine a large tribe of Chumash here; the men hunting rabbits in the dunes, the beautiful baskets of the women overflowing with shellfish.

Continue down to the ocean beach side of the spit and hike another 2½ miles. Approximately 1½ miles from the north end of the spit, in an area of little vegetation, climb over the dunes to the east toward Morro Bay. Here you'll find sheltered Houseboat Cove. Across the bay you'll spot the Morro Bay State Park Museum.

Walk along bay shore. Many birds, such as sandpipers and curlews feed in the mudflats. The vegetation increases, with saltgrass and buckwheat predominating. Continue hiking northeast along the shore until reaching the north end of the spit. Catch the Clam Taxi, or return the same way.

Black Mountain Trail (SLO-6)

Picnic Ground to Peak; 3 mile RT; 600′ gain

A series of nine peaks between San Luis Obispo and Morro Bay originated as volcanoes beneath the sea that covered this area 15 million years ago. After the sea and volcanic explosions subsided, erosion began dissolving the softer mountains material around the volcanic rock and left the nine volcanic peaks standing high above the surrounding landscape. These volcanic plugs include Islay Peak in Montana De Oro State Park, Hollister Peak, and the famed Morro Rock.

Black Mountain, the last peak in the volcanic series before Morro Rock, has a trail that tours through a little of everything—chaparral, eucalyptus, oaks, pines, and coastal shrubs. From the mountain's 640-foot summit, you can see the Morro Bay Estuary, the sandspit, and the hills of nearby Montana De Oro State Park.

Directions to trailhead: Follow Highway 1 twelve miles north of San Luis Obispo to the Los Osos-Baywood Park exit just before Morro Bay. Turn south on South Bay Boulevard and go ¾ mile to Morro Bay State Park entrance. Bear left on the first fork beyond the entrance, heading ¾ mile to the campground entrance. Park along the first crossroad inside the campground. Walk up the campground road to the picnic ground, where you'll see a pipe gate which indicates the beginning of the trail.

The Hike: Follow the exercise trail, cross a paved road and begin ascending more steeply. A mile from the trailhead, there's a junction. Bear left. The route becomes steeper, passing first through coastal shrubs, then conifers. The trail passes a water tank, then switchbacks to the summit.

After enjoying the fine view, you may return the same way or follow the option below.

Option: Return via East Fork of Exercise Trail: After you backtrack the ½ mile to the trail junction, go straight, to the east. You'll discover a eucalyptus grove, where monarch butterflies cluster. Cross a golf course road and rejoin the eastern section of the exercise trail, which returns you to the trailhead.

Leffingwell Landing Trail (SLO-7)

Shamel County Park to Leffingwell Landing; 2.5 miles RT
Shamel County Park to Moonstone Beach Drive; 4 miles RT

Named for moonstones (a milky translucent agate), gravelly-shored Moonstone Beach is located along Moonstone Drive in Northern Cambria Pines. Determined rock hounds occasionally discover bits of jade amongst the piles of driftwood along this beach. In January and February, gray-whale watching is excellent here because the giants swim so close to shore.

This is an excellent hike for observing otters and tidepools. The trail follows San Simeon State Beach and visits Leffingwell Landing, the site of a pier once figuring prominently in 19th century coastal trade and now a fine picnic area.

Directions to trailhead: Follow Highway 1 to Cambria, 21 miles north of Morro Bay. Turn left on Moonstone Beach Drive, bearing left across a bridge onto Windsor Boulevard. Park at Shamel County Park.

The Hike: From the county park, walk north along Moonstone Beach, which at this point is composed of colored pebbles instead of sand. In a quarter mile, you'll ford Santa Rosa Creek. The creek mouth is sometimes difficult to cross at high tide and after winter storms. The beach narrows. A little less than a mile from the trailhead, climb the low bluffs and begin following a series of unmarked trails through coastal sage atop the bluffs. You may see sea lions sunning themselves at a rocky cove below.

When you reach the old highway bridge, cross the usually dry creek. Just above the creek is the picnic area at Leffingwell landing, nestled in a sheltered cypress grove. During the 1870s and 1880s, ships unloaded lumber and other goods here for the pioneer settlers along San Simeon Creek.

Option: to Moonstone Beach Drive—The trail picks up again at the bluff edge past the picnic area. Soon you'll be treated to views of Piedras Blancas lighthouse to the north and Hearst Castle inland. Indian mortar holes are ground into the tops of the sandstone bluffs rising above the beach. The trail passes behind a grove of cypress trees and soon descends to the beach. Moonstone Beach Drive is a short rock-scramble inland.

Return the same way or via Moonstone Beach Drive.

Monterey County

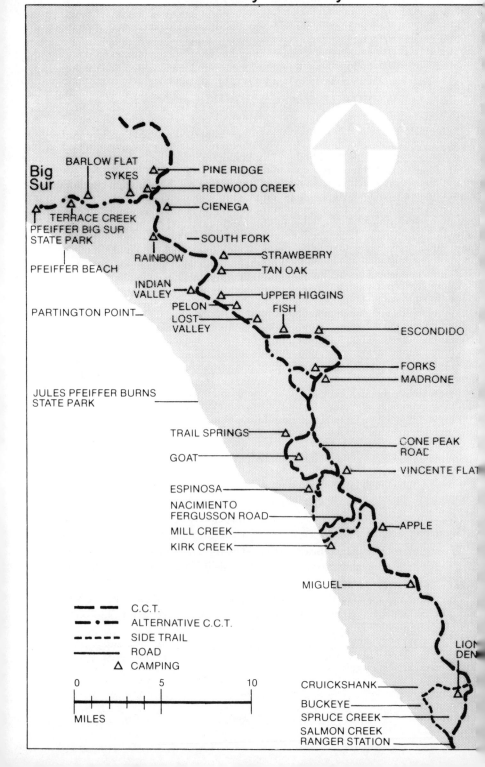

Monterey County (south)

Monterey County's coastline extends from the dramatic Big Sur area to the flat sandy coastal plain along Monterey Bay. The city of Monterey is rich in history. Here in 1770, Father Serra established a presidio and the second California mission. Under Mexican and Spanish rule, the city was the capital of Alta, California. In Volume II of this guide, CCT will visit Monterey, as well as Pt. Lobos, Carmel-By-The-Sea, Moss Landing, and many more beautiful and historic spots in the northern part of the county.

This volume explores the southern half of the county from the San Luis Obispo County line to Pfeiffer Big Sur State Park. Both geographically and spiritually, Big Sur is the heart of California. CCT travels through this heartland across the rugged mountainous terrain of the 159,000 acre Ventana Wilderness. From the windswept ridges of the Santa Lucia Mountains, hikers look down at what is often called, "the greatest meeting of land and sea in the world."

CCT ambles among the headwaters of the Arroyo Seco, Little Sur and Big Sur Rivers, which originate in the Ventana Wilderness. Observant hikers may spot a rare and beautiful spire-like tree, the Santa Lucia Fir, or bristlecone fir, which grow only in the Santa Lucia Mountains. In these mountains also is the southern-most limit of the natural range of the coast redwood. Fern-lined canyons, oak studded potreros and meadows smothered with Douglas' iris, pink owl's clover and California poppies welcome the backcountry traveler.

Hikers proclaim the last 100 miles of trail described in this guide as the most glorious section of the California Coastal Trail. Big Sur country is not a gentle wilderness, but it is a dramatic, enchanted land; a fitting climax to the southern half of the trail and an invitation that cannot be refused to the second half.

CCT at a glance

TERRAIN: Mountain trail from Salmon Creek to Pfeiffer Big Sur State Park. Trail follows both creek drainages on the western slope of the Santa Lucia Mountains and fire roads and trails on the ridge crest. Cool, redwood-lined canyons, oak woodland and chaparral are some of the diverse natural communities traveled by CCT.

OBSTRUCTIONS: A nasty few miles of highway walking is required from the San Luis Obispo County line to the first entrance of the Los Padres National Forest at Salmon Creek. Trail maintenance is spotty throughout the Ventana Wilderness. No bridges in the Wilderness; river crossings difficult during wet season.

TRANSPORTATION: Monterey Peninsula Transit: 1 Ryan Road, Monterey 93940, (408) 899-2555 or 424-7695. Coastlines Bus Service: (408) 649-4700.

CAMPGROUNDS/ACCOMMODATIONS:
Coastal camps (not along CCT) oriented toward vehicles, include:
Plaskett Creek
Kirk Creek
Limekiln Beach Redwoods (private)
Pfeiffer Big Sur State Park
CCT hikers will find plenty of trail camps enroute to the Los Padres backcountry

WILDERNESS PERMITS are required for travel in the Ventana Wilderness, and can be obtained from the Monterey District Ranger's Office: U.S. Forest Service, 406 S. Mildred St., King City 93930; (408) 385-5434. Four other Forest Service Stations along the coast offer information, maps, and permits. Big Sur Station is located off Highway 1, .7 mile south of Pfeiffer Big Sur State Park; Bottcher's Gap Station is 8 miles up Palo Colorado Canyon Rd.; Pacific Valley Station is 33 miles south of Big Sur Valley; Salmon Creek Station is 45 miles south of Big Sur Valley and a few miles north of the San Luis Obispo County line. Call (408) 667-2423.

CCT: MONTEREY COUNTY

In 1978 the Marble Cone Fire extensively damaged the Los Padres backcountry. Much of what was forest is now brushland and may not regain its timbered beauty for over a hundred years. During the first few years after a major fire, undergrowth regenerates at a phenomenal rate as the natural succession of the forest begins another cycle. In the Big Sur country south of the state park, many of the ridges were scorched, although a number of steep timbered canyons and watersheds escaped devastation. The trail system has suffered both from erosion and from a profusion of brush, and has received little maintenance since the fire. Deadfalls and washouts are common. Trails are distinguishable because the tread still exists, but many miles are overgrown, compelling the hiker to constantly push brush out of the way to continue. Before undertaking a lengthy trip, hikers should inquire about trail conditions at a ranger station.

CCT re-enters the Los Padres National Forest a hundred yards south of Salmon Creek Ranger Station. Our trail follows the signed Salmon Creek Trail, beginning on the east side of Highway 1, on the south side of the creek.

CCT/Salmon Creek Trail immediately begins climbing, first through lush streamside vegetation, then across the exposed slopes of the canyon, which are covered with seasonal wildflowers. Lingering summer fog seems to protect flowers here; unlike their more southern cousins, they survive into early summer, testifying to the truth of the immortal words of British poet Oliver Goldsmith: "Flowers grow best where broad ocean leans against the land."

A thousand feet above sea level, the trail crosses a stream and ascends into a forest of Douglas fir, often called Spruce—which helps explain our forthcoming destinations of Spruce Creek and Spruce Camp.

Two miles from the trailhead is the Spruce Creek Trail junction. The trail to the right leads south toward Dutra Spring and Carpojo Creek. A few hundred yards of hiking and you'll drop down to Spruce Creek Camp, located at the confluence of the waters of Salmon Creek and Spruce Creek. Spruce Creek Camp is in the deep shade of, uh, firs.

(If you're good at boulder-hopping and the creek isn't too high, it's possible to pick your way up Salmon Creek a mile to Estrella Camp. Remains of hydraulic mining equipment will be spotted enroute.)

Our trail resumes on the other side of Spruce Creek and continues along the south slope of Salmon Creek. You cross a meadow, where a cabin, believed to have been occupied by one of the hydraulic miners, once stood. Only the crumbling stone foundation remains. The trail continues ascending moderately to Estrella Camp (1500'), a grassy shady area along Salmon Creek. From here to Coast Ridge Road is no dependable source of water, so fill up here. The trail soon rises above the last trees and ventures out onto the hot, brushy upper canyon slopes. You climb 1800' in the next 2½ miles through an eroded area that is just now recovering from a serious 1970 fire. This ia very hot stretch of trail in summer!

CCT arrives at Coast Ridge Road (3120'), a fire and military road marking the boundary between Fort Hunter Liggett Military Reservation and the National Forest. CCT will be following this road for approximately the next 20 miles. Traffic will pose little hassle for the hiker, except for summer dust. What *is* a problem is the lack of water. Coast Ridge Road stays high on the coast ridge, and ridgetops, as experienced hikers know, rarely have water. Camps and water are generally on the west slope of the ridge and require a steep descent.

Bear left on the road. On clear days you'll be able to see the ocean to the west, Salinas Valley to the east. In .1 mile hikers reach the signed junction of the Cruickshank Trail. (A sign indicates it's a mile to Lion's Den Camp, but it's more like ½ mile.) A brief descent along a rough eroded jeep road brings you to the camp, a small flat area often perched just above coastal clouds. Water is supplied by a small creek.

CCT ascends with Coast Ridge Road, climbing 500 feet in the next four miles past Alder Peak. In another ½ mile you'll come to a signed junction. (A left turn will take you on a 1¾ mile descent to another junction. Alder Creek Camp [2240'] is a 1½ mile descent to your left.) The camp, which is accessible to those car campers willing to brave the rugged road up from Highway 1 has three stoves and plenty of tenting space amidst oaks on the banks of dependable Alder Creek. A right turn from the previous junction takes hikers six miles down to Pacific Coast Highway just south of Willow Creek Picnic Area.

CCT continues on the serpentine Coast Ridge Road, ascending and descending with the steep rocky jeep road for five more miles to the signed junction with Miguel Camp. The camp is ¼ mile from the road. One stove, one table. No water.

Continuing with Coast Ridge Road for two more miles hikers reach the unsigned junction with Plaskett Ridge Road. A left turn takes hikers on a 1½ mile descent to Plaskett Ridge Camp, a small camp with one table, one stove, and *no* water. One mile north of Plaskett Ridge Camp turnoff, CCT arrives at another junction; the steep road to the right continues along the national forest boundary. We bear left to yet another junction, this one with Prewitt Ridge Road.

If you bear left on Prewitt Ridge Road in ¼ mile you'll see a camp, 100 yards from the road on your left. One stove. A short distance farther on the road is a signed junction with an arrow pointing to Prewitt Ridge Camp, a shady spot with a stove near some handsome boulders. No water is at either camp.

If you wish, descend sharply with Prewitt Ridge Road through mixed oak and conifer forest. A few clearings offer sweeping vistas of the coast. The lovely road drops down through madrone, yerba buena, and scads of gooseberries for 2½ miles to road's end at Alm's Ridge. The recreation site indicated on USFS maps no longer exists. However, a developed spring is here, surrounded by a stock-trampled mudbank.

In theory, a Forest Service "infrequently maintained" trail leaves Alms Ridge, heading east then north for about ¾ mile to Mill Creek Trail. What appears to be the correct trail leaves from Alms Ridge Spring, contouring along a cool redwood slope for a few hundred yards and promptly dead-ending in an abandoned stock enclosure. Other trails in the vicinity appear to be ancient cattle or deer trails. No doubt, you'll end up striking cross-country on a steep northerly descent until you reach Mill Creek, which is paralleled by Mill Creek Trail. (See Mill Creek Trail description.) This trail leads east to Nacimiento Summit and west toward Pacific Coast Highway.

Cone Peak Trail

CCT continues with Coast Ridge Road, dipping down and up for 3¾ miles to Apple Camp, a not-very-exciting spot with one table, one stove, no water, no apples. Descending another two miles you reach the intersection with the Mill Creek Trail on your left and dead ahead, Nacimiento Summit. CCT crosses Nacimiento-Fergusson Road and begins ascending steeply up the signed Coast Ridge/Cone Peak Road. In 3½ miles you'll reach the signed junction with the Girard Trail. (It's another 2½ miles of stiff climbing to the end of Coast Ridge Road and the start of Coast Ridge Trail just below Cone Peak.) You'll find the Girard Trail junction when the road comes to a saddle.

Until a few years ago, the Girard Trail was officially abandoned, but the Young Adult Conservation Corps did a wonderful job of resurrecting it. CCT/Girard climbs west over a ridge for ¼ mile, briefly levels off, the descends abruptly. If you look carefully, you might spot the remains of the old Girard Trail, which once followed this ridge all the way to Highway 1 and Limekiln Creek.

After several sharp switchbacks down a dark shady hillside, the trail reaches a quiet grove of redwoods with a stream trickling down from the north. You'll pass a small flat campsite, not indicated on USFS maps. The trail follows a stream, crossing it several times. CCT arrives at Vincente Flat Camp. At the lower end of the campground, you'll find the Stone Ridge Trail, which will take you north toward Goat Camp.

The trail sign indicating it's 3½ miles to Goat Camp should not be believed. Maybe, if you're a goat, it's correct. Two-legged animals will find it's more like five. CCT ascends gradually, contouring out to western slopes where there are fine views of the ocean. The trail ascends into Limekiln Creek drainage and crosses the creek. As you hike the upper reaches of the canyon, the trail becomes difficult to follow because range cows have trampled the main trail into oblivion and added a few paths of their own. Stick with the left side of the drainage and follow the colored plastic ribbons attached to tree limbs. When you near the top of the drainage, the trail grows more distinct. From the ridgetop, there's a fabulous view of the Pacific and the wooded watershed of Limekiln creek. The trail heads north and arrives at Goat Camp (2240'), which occupies a small grassy ridge between two dependable creeks. The camp has a table and grill.

CCT continues north and east out of Goat Camp on a series of very steep switchbacks, crossing chaparral and grassland. A mile from Goat is the signed Ojito Camp Junction (3540'). The Stone Ridge Trail we've been following plummets 700 feet in ½ mile to Ojito Trail Camp, an attractive campsite streamside among madrone. Lillian Bos Ross wrote of "los ojitos" in her novel about Big Sur, *The Stranger*. At a wedding celebration in the mountains, Bill the fiddler is asked to play "Dance of the Little Bears," and struggles to pronounce it in Spanish.

"No, not 'los ojitos'—that's saying 'the little eyes.' What you want to say is 'los ositos,' when you mean 'the little bears.'"

Avery laughed good-naturedly and said after Mel. "Ojitos-Ositos-eyes, bears. Not much difference to the sound but a lot of difference to the way you'd dance it."

CCT ascends from the Ojito Camp Junction on the well-maintained Gamboa Trail through sugar pine for 1¾ miles, arriving at Trail Springs Camp (3800'). This is a steep, somewhat dark and gloomy camp with a stove near a small spring.

From Trail Spring Camp, CCT leads eastward through woods, ascending 1¼ mile to the signed junction with Coast Ridge Trail (4600'). A right at the junction takes hikers on a 1½ mile descent through sugar and ponderosa pine to the Cone Peak Roadhead.

CCT takes hikers to the left along Coast Ridge Trail (which can be followed by the intrepid all the way to Ventana Inn Road Junction at Highway 1). CCT contours a mile through pines with little elevation change to a signed junction. A sidetrail leads ¼ mile to Cook Camp.

Coast Ridge Trail/CCT soon becomes a road, though years of erosion and encroaching brush make it seem like a trail. Sometimes you'll travel under a canopy of brush, sometimes you'll break into the open for terrific views of the coast. Arroyo Seco Trail junction comes up in a mile. A sign indicates it's two miles to Madrone Camp.

At this junction, CCT hikers may choose from two options: Continuing straight ahead on Coast Ridge Road (3F10) is the most direct route northward. As its name suggests, it sits squarely atop the ridge, marching steeply up and down the spine of the crest. It's hot, dry, waterless and a number of unsigned dozer trails confuse matters. Additional dozer trails left over from the Army's use of this area as a tank proving ground also confuse the hiker. Stay on Coast Ridge Road for 5.3 miles to the unsigned junction with an old tractor trail (4E06). Drop east from the ridge on this steep, overgrown rocky path. The switchbackless dozer trail is a knee-killer. At the intersection with Higgins Creek Trail, you'll meet the other CCT route coming in from the east from Fish Camp.

CCT's second option is a half day's travel longer than the route along Coast Ridge, but stays mostly on foot trail. It makes a lazy half circle east to the Arroyo Seco River and then north and west over to Lost Valley.

At the signed Arroyo Seco Trail Junction on Coast Ridge Road, bear right. CCT/Arroyo Seco Trail descends rapidly through madrone, oaks, scrub oak, manzanita and knob cone pine. Descending with the trail are remnants of an old field phone line used during World War II. The trail reaches the Arroyo Seco River, then follows it downstream a short ways to Madrone Camp (2855'). The camp is secluded in a shady spot above the river. One stove, one fire ring.

From Madrone, the trail continues downhill through the redwood lined canyon. More traces of the Army's abandoned field phone line can be seen. You descend to your right, crossing and re-crossing the Arroyo Seco. The trail continues a short ways through woodland, crosses a stream, and shortly becomes a tractor trail. Another tractor trail, unsigned, descending from the northwest intersects our route. This tractor trail (4E26) sometimes known as the Rodeo Flat Trail makes a brutal ascent a few miles up to Coast Ridge Road, and adds another option to the northward progress of CCT.

Continuing on the Arroyo Seco Trail for another hundred yards brings you to Forks Camp (2430') bordered by Forks Creek and the Arroyo Seco. One stove, plenty of flat tenting space.

A bit beyond camp, the tractor trail gives way to foot trail once more. We follow the banks of the Arroyo Seco. In cooler places, cedars and Santa Lucia fir grow along the river. One and three quarters miles from Fork Camp, you reach a dirt road and follow it briefly through some private property belonging to the Southern Monterey County Sportsmen's Association. You'll soon reach a signed three-way intersection on Indians Road. You will bear northwest at this intersection toward Escondido Camp, 2½ miles away. The dirt road climbs gently along the east bank of the Arroyo Seco River.

Escondido Camp (2170) is a handsome relatively undeveloped auto campsite. Piped water, nine campsites, swarms of insects in summer.

CCT follows the camp road for ⅛ mile to the signed trailhead of the Higgins Creek Trail. This trail descends for a mile, losing 500' in elevation down to the Arroyo Seco River. Crossing the river, the trail begins a vigorous ascent, for a mile sticking close to a stream, then heading into hot, exposed brush-covered slopes. The trail climbs up narrow trail, where an exceptionally large number of ticks await hikers, to the divide (2800') separating Arroyo Seco and Lost Valley watersheds. Atop the divide is a burnt post, the charred remains of a trail sign wiped out in the 1977 fire. The trail descends into Lost Valley watershed, passing a few dry streams and an occasional wet one. A mile from the divide summit, the trail crosses Fish Creek and arrives at Fish Camp. Two stoves, plenty of tent space.

Leaving camp, CCT ascends through grassland and around a pine slope and in 1¼ miles reaches a saddle with a superb view of Lost Valley. Here a sign indicates it's one mile to Lost Valley Camp. Next to the sign you'll notice the awful tractor trail (4E09) coming in on your left down from Coast Ridge Road. Those hikers selecting the Coast Ridge Road option will join us at this junction. Arguments will ensue as to which route is the more difficult.

From the saddle the trail descends ¾ mile to Upper Lost Valley Camp (not on USFS map). It's a tiny campsite across from Lost Valley Creek.

CCT continues another ¼ mile along the creek to the main Lost Valley Camp, circled by towering pines. Several stoves, reliable water, and popular with the pack horse gang.

Leaving Lost Valley Camp, CCT contours northwest, crossing Lost Valley Creek and passing another tractor trail (not shown on USFS maps) which ascends steeply up to Coast Ridge Road. CCT enters an area of burned pine forest, climbing over a small hill and crossing Higgins Creek. Obscured by brush is a plaque in a boulder which reads:

> "In rembrance of Bill Cotta and his horse who lost their lives in the raging waters of this river near this point on the thirty-first day of January 1963. This plaque was placed here lest men forget the awesome forces of nature."

During periods of high water, crossing this creek can be treacherous. A word to the wise...

CCT makes several crossings of the creek. It may be difficult locating some of the crossings after a rainstorm. The trail switchbacks briefly uphill, then descends to Pelon Camp (2,000′). Located on a small rise above Higgins Creek, this small campsite stands forlornly among trees lost to fire. In short order, CCT crosses Pelon Creek, then Higgins Creek, and reaches another Pelon Camp, this one shadeless with a single stove.

Leaving Pelon, the trail follows Higgins Creek for a mile to Upper Higgins Camp, a spacious open site with a stove. Leaving the camp, CCT crosses the creek and begins switchbacking uphill through mixed pine and oak forest. We emerge on a low ridge, then return to Higgins Creek soon reaching and crossing Indian Valley Creek. CCT soon arrives at a junction where a sign points the way to Indian Valley (to the right) and that's where we head. We climb briefly then drop to the camp by Indian Valley Creek. Surrounded by pines, the meadowland camp is pleasant, but the creek is not to be counted on during the dry season.

CCT continues along with the west fork of the Higgins Creek Trail upstream to (Old) Indian Valley Camp (2800′) at the junction with the Marble Peak Trail. One stove, undependable water. A trail to the left leads to Marble Peak. CCT bears right, northeast, ascending through scorched pine forest and sandstone outcroppings. CCT continues to a summit ridge, then descends rapidly along the divide between Higgins Creek and Zig-Zag Creek watersheds. CCT comes to a saddle, then descends to Tan Oak Creek, following an old tractor trail. This tractor trail is extremely overgrown, first with yucca on its upper end, and ferns on its lower end. The tractor trail reaches the creek, where we intersect an extremely overgrown trail and follow it along Zig-Zag Creek for a brief time to Tan Oak Camp (2680′). No stove, water and wood plentiful. The camp is in deep shade.

Leaving Tan Oak Camp, CCT heads north, following Zig Zag Creek, and ascending gently to the signed South Fork Trail Junction. CCT heads right toward Strawberry Valley, passing through a barbed wire fence, once used as a corral. The trail ascends moderately for half a mile through Strawberry Valley to Strawberry Valley Camp. Sunny, roomy, two stoves, intermittent water in summer.

CCT follows the South Fork Trail west from camp, climbing a low rise to the divide between the Arroyo Seco and Big Sur watersheds. In 1½ miles you reach South Fork Camp (1800') located just upstream from the confluence of the Big Sur River and Pick Creek. It's a flat sunny spot amongst old oaks. Pick Creek is a fun exploration; in times of high water, you'll discover a waterfall and a circular pool.

CCT crosses the Big Sur River several times, before sticking with the west bank for a spell and following it downriver for 3½ miles to Rainbow Camp (1560'). Just below camp is a swimming hole and a small sandy beach. Rainbow has two stoves and plenty of tent spaces.

(The de Angulo Trail leaves this camp, heading for Mocho Camp and Highway 1, 8¾ miles away.)

CCT departs the camp from the east side of the South Fork of the Big Sur River and begins some steep switchbacks. CCT continues climbing over a thousand feet to a saddle atop the ridgeline between the north and south forks of the Big Sur River. CCT then switchbacks down to the north fork of the Big Sur River and crosses it. Downstream from the river crossing, at the confluence of Cienega Creek and the north fork of the Big Sur River, are some especially captivating pools and waterfalls.

The trail climbs again for a mile to Cienega Camp (1800') located on dependable Cienega Creek. Campsites are rocky. Rumor has it a hot spring is located a short ways upstream.

Leaving camp, CCT veers westward, crossing a cienega (Spanish for swamp). The first 200 yards of trail after crossing Cienega Creek is incredibly overgrown with ceanothus. Beyond the jungle the trail continues ascending steeply, but the oak copses on the higher slopes are easier traveling. The trail reaches a saddle, descends briefly and reaches a signed junction with the Pine Ridge Trail; this trail leads east to Pine Ridge Camp, as well as west toward Redwood Creek.

The Pine Ridge Trail Junction is as far north as CCT proceeds in this volume. Volume II describes CCT north through the Los Padres, out to the coast, and, in fact, all the way to Oregon.

The trail to Pfeiffer Big Sur State Park, a distance of 13 miles, is briefly described below:

Turn west on the Pine Ridge Trail, descending moderately for one mile through oaks to Redwood Creek Camp (1800') which has several stoves and tent sites on both sides of Redwood Creek. It's a lovely camp in deep shade and a rarely visited alternative to overused Sykes Camp farther down the trail. West of camp is the site of an ancient Indian camp, where mortar holes can be found in the rocks.

Leaving camp, the trail follows the north side of Redwood Creek along chaparral slopes, descending to Sykes Camp (1080') on the east bank of the Big Sur River. This camp has a little bit of everything—morning sun, afternoon shade, a deep swimming hole and a hot springs. The hot springs is ¼ mile down river and flows south of the riverban. You'll enjoy basking in the sulphurous 100° waters, gazing up at the stars in the night sky. Only ten miles from Pfeiffer Big Sur State Park, Sykes Camp is a (too) popular destination. Most visitors are of the mellow variety, but others lack wilderness manners and leave litter behind. Drinking water taken from the river should be purified. Syke's charms are undeniable, but seekers of solitude should steer clear in the summer months. Rangers report that nearly 80 percent of the backcountry use taking place in the Monterey District of the Los Padres Forest occurs along the ten miles from the state park to Sykes!

Departing Sykes, the trail crosses the river, which can rise to dangerous levels during the rainy season. The trail ascends a ridge, staying above the river until it reaches Barlow Flat, a flat expansive camp on the north side of the river in the shade of redwoods. It's especially popular with fishermen. Purify drinking water taken from the creek.

The trail soon crosses Logwood Creek, climbs a hill, descends through woodland, and in a mile reaches Terrace Creek. A short side trail, Terrace Creek Trail, leads up-creek to Terrace Creek Camp, shaded by redwoods. Because it's so close to the trailhead, it's apt to be crowded.

Our trail crosses Terrace Creek, leading westward through woodland, passing the junction of the trail to Ventana Camp (1¼ miles north). From this junction we continue dropping gently westward. Mount Manuel dominates the northern skyline and hikers can see the Pacific and down into Big Sur Valley. (It's possible to improvise a short cross-country route down to Pfeiffer Big Sur State Park campgrounds.) The trail continues descending through woodland, then chaparral, ending abruptly at the National Forest parking lot near the Big Sur Ranger Station.

Salmon Creek Trail (M-1)

Salmon Creek-Cruickshank-Buckeye Loop; 14 miles RT; 3000' gain

This hike, suitable for a strenuous day hike or more leisurely weekend backpack, offers a chance to sample the diversity of the coast range—lush fern canyons, fir forests, oak potreros—and sweeping views of the majestic coast and the Salinas Valley.

The lower part of the Salmon Creek canyon sees heavy use with a fair amount of illegal camping; upper reaches are far less visited. Steelhead used to run here.

Directions to trailhead: Salmon Creek Ranger Station is located a few miles north of the Monterey/San Luis Obispo County line off Highway 1. For those journeying from the north, the forest service station is 76 miles south of Monterey. The trailhead is a hundred yards south of station. Ample offroad parking at the station. The signed Salmon Creek Trail begins on the east side of the highway on the south side of the creek.

The Hike: Salmon Creek Trail immediately begins climbing, first through lush streamside vegetation, then across the exposed slopes of the canyon, covered with seasonal wildflowers. The lingering summer fog seems to protect the flowers here and their display.

A thousand feet above sea level, the trail crosses a stream and ascends into a forest of Douglas fir, often called Spruce—which helps explain our forthcoming destinations of Spruce Creek and Spruce Camp.

Two miles from the trailhead is the Spruce Creek Trail junction. The trail to the right leads south toward Dutra Spring and San Carpojo Creek. The main trail continues straight ahead up the main canyon of Salmon Creek. A few hundred yards of hiking and you'll drop down to Spruce Creek Camp, located at the confluence of the waters of Salmon Creek and Spruce Creek. Spruce Creek Camp is in deep shade.

(If you're good at boulder-hopping and the creek isn't too high, it's possible to pick your way up Salmon Creek a mile to Estrella Camp. Remains of hydraulic mining equipment will be spotted en route.)

Our trail resumes on the other side of Spruce Creek and continues along the south slope of Salmon Creek. You cross a meadow, where a cabin, believed to have been occupied by one of the hydraulic miners, once stood. The trail continues ascending moderately to Estrella Camp (1500'), a grassy shady area along Salmon Creek. From here to Coast Ridge Road, there's no dependable source of water, so fill up here. The trail soon rises above the last trees and ventures out onto the hot, brushy upper canyon slopes. You climb 1800' in the next 2½ miles through an eroded area that is just recovering from a serious 1970 fire. This is a very hot stretch of trail in summer!

You reach the high point of the trail at Coast Ridge Road (3120'), a fire and military road marking the boundary between Fort Hunter Liggett Military Reservation and the National Forest. You will bear left on the road. On clear days you'll be able to see the ocean to the west, the Salinas Valley to the east. In .1 mile you'll reach the junction of the Cruickshank Trail The sign indicates it's a mile to Lion's Den Camp, but it's more like ½ mile. A brief descent along a rough eroded road brings you to Lion's Den Camp, two small flat areas, often situated just above the coastal clouds. Water supply is from a small creek.

Leaving Lion's Den, you follow Silver Peak Road ½ mile to a junction. (Peak-baggers won't overlook Silver Peak [3590'] on the left.) Cross the road and follow the Cruickshank Trail. The trail descends, crossing a creek, and drops 1000' in the next 2½ miles. You'll get fine views of the Villa Creek drainage; in spring, waterfalls can be seen cascading down the canyon. Silver Camp, not shown on USFS maps, is a streamside camp with one stove and plenty of flat tenting sites.

Three-quarters of a mile from Silver Camp, we veer south on the Buckeye Trail. We begin ascending through heavy timber, climbing the shady north slope. The trail descends to Redwood Creek, crossing it and proceeding in a southerly direction along the ridge separating Villa Creek and Redwood Creek Canyons. The trail grows more tentative as it enters a meadow and reaches Buckeye Camp (2140'). There's one campsite on the near side of the meadow, another on the far side next to a stock enclosure. The latter camp has a developed spring.

Leaving the meadowland, you contour around to the western slopes, receiving the twin pleasures of ocean breezes and coastal views. You descend a ridge, cross Soda Springs Creek, and arrive at a signed junction (500'). The Buckeye Trail continues on to the coast highway. Our trail, which completes the loop heads south. This trail, not on USFS maps, descends a mile through grassland and chaparral back to the Salmon Creek Ranger Station.

Vincente Flat Trail (M-2)

Highway 1 to Espinosa Camp; 6.6 miles RT; 1600′ gain
Highway 1 to Vincente Flat Camp; 10 miles RT; 1600′ gain

Vincente Flat Trail provides an ideal introduction to the charms of Big Sur, for in five miles the hiker experiences meadowland, coastal and canyon views, and a redwood forest. Vincente Flat Camp is an ideal picnic site.

Hikers wishing to join CCT may do so by taking this trail to Vincente Camp.

Directions to trailhead: The signed Vincente Flat Trailhead is located opposite Kirk Creek Campground on Highway 1, just north of the Nacimiento-Fergusson Road turnoff.

The Hike: The trail immediately begins ascending on a series of well-graded switchbacks through brush and grassland. Sweeping views of the coast from Jade Cove to Gamboa Point are yours. One nice feature about this trail is the way it alternates from sunny exposed slopes to shady redwood ravines. After crossing over a ridge, the highpoint of the Vincente Flat Trail, and enjoying the fine coastal view, you'll enter the watershed of Hare Canyon.

Tiny Espinosa Camp (1780′) is 3¼ miles from the trailhead. Water is ¼ mile up trail, where a tiny unnamed creek cascades down a redwood-lined ravine to the trail.

The trail ascends briefly, then makes a short descent to Hare Creek, follows the creek, then crosses it. On a low rise above the creek is the signed Stone Ridge Trail Junction. To reach Vincente Flat Campground (1600′), take the Girard Trail to the right upstream 150 yards. Redwoods shade idyllic campsites. Plenty of water flows, even in summer, and a lovely meadow beckons sun worshippers and frisbee flyers.

Return the same way.

De Angulo Trail (M-3)

Highway 1 to Coast Ridge Road; 7 miles RT; 2700′ gain

Jaime de Angulo was one of the most eccentric of the rugged individuals who've called Big Sur home. In 1914, he built an unusual house perched 2,000 feet above the ocean atop Partington Ridge. Scandalized neighbors reported that he was frequently seen outside plowing, clothed only with a red neckerchief about his head. A one-time doctor, de Angulo studied Indian lore and wrote a book on their mythology called *Indian Tales*. He cooked Indian style over a fire ring on his living room floor, chopping a hole in his roof to allow smoke to escape.

The de Angulo Trail leads up Partington Ridge through grassland, tan oak woodland, chaparral, and redwood forest to Coast Ridge Road, where sweeping coastal vistas greet the determined hiker.

Because of the heat and exposed terrain, as well as the stiff elevation gain, this trail is not recommended for travel during summer months. Spring brings fine wildflower displays and blooming yucca and a clear blustery fall day rewards the hiker with terrific views. A second word of warning: Although Jaime de Angulo used to put signs on his property reading "Trespassers Welcome," the current residents of Partington Ridge are not so accommodating. This trail passes near several residences. Respect private property.

Hikers wishing to join CCT may follow the de Angulo Trail. From Coast Ridge Road, the trail descends to Cold Spring Camp and Mocho Camp, intersecting CCT at Rainbow Camp, 9 miles from Highway 1.

Directions to trailhead: Unsigned de Angulo Trail departs the east side of Highway 1 at a turnout 8 miles south of Peiffer Big Sur State Park and about 1 mile south of Torre Canyon bridge.

The Hike: Hike up the private dirt road that ascends the hill above the parking area. 200 feet up the road, a sign reading "trail" points to a northbound path. The trail ascends into brush, soon passing through a gate and entering pasture land. A profusion of stock trails make it difficult to follow. A few small "trail" signs with arrows help out. At the second of these signs, be sure to veer left (north) around a huge storm-wrecked oak, then along a fence line.

The trail meets a private road, following it north. When the road forks, take the upper road, where a wrecked car overlooks the junction. Soon yet another trail sign indicates a northbound path. Follow it through a thicket of sweet peas, domestic plants thriving in the wild. The trail passes into woods and by a number of old wrecked cars and trucks.

Ascending through tan oak woodland and redwood forest, the trail re-emerges onto sunbaked grassland and comes to a signed tractor trail. Follow this trail very briefly to another sign indicating the trail leads upward. The trail climbs through grassland and oak copses and through a stand of pine atop the ridge separating Partington and Torre Canyons. Climbing through brush, the trail intersects a steep boulder-strewn firebreak and follows it upward. The trail leaves the firebreak and contours gently to the north over to Coast Ridge Road. Superb views to the east and west and fine picnic spots reward the hiker.

Return the same way.

Pfeiffer Beach Trail (M-4)

Sycamore Canyon Road to Pfeiffer Beach; 1 mile RT

Los Padres is one of only three National Forests in America with ocean frontage. Named for the pioneer Pfeiffer family, this secluded white sand beach faces the turbulent sea which sends awesome waves crashing through blowholes in the rocks. Many scenes from the film "The Sandpipers," with Elizabeth Taylor and Richard Burton, were shot here. The magic of motion pictures gave us a calm beach where small boats easily landed. Big Sur residents laugh every time this movie is shown on late-night TV.

With its hazardous surf and gusty winds, Pfeiffer Beach cannot be said to be a comfortable stretch of coastline; it is, however a magnificent one.

Directions to trailhead: Driving south, a mile south of Big Sur State Park entrance, take the second right hand turn off Highway 1 (west). Sycamore Canyon Road is a sharp downhill turn. Follow the two mile narrow, winding and sometimes washed out road to the Forest Service Parking area.

The Hike: Follow the wide sandy trail leading from the parking lot through the cypress trees. Sycamore Creek empties into a small lagoon near the beach. Marvel at the sea stacks, blowholes and caves and try to find a place out of the wind to eat your lunch. The more ambitious may pick their way over rocks northward for a mile around a point to a second crescent-shaped beach.

Tanbark Trail (M-5)

Highway 1 to Partington Cove; ½ mile RT; 200' loss

Partington Cove

Partington Cove, part of Julia Pfeiffer Burns State Park, was once the site of a dock where tanbark was loaded onto waiting ships. Woodsmen stripped the tanbark oak, a kind of cross between an oak and chestnut. Before synthetic chemicals were invented to tan leather, gathering and shipping of the bark was a considerable industry in these parts.

This short leg-stretcher of a hike, drops down to Partington Creek and over to the deep blue waters of the cove.

Directions to trailhead: 1.8 miles north of Julia Pfeiffer Burns State Park entrance, look for an iron gate on the west side of the road. A turnout for parking is around the next bend on the west side of the road.

The Hike: From the iron gate, a dirt road drops down to Partington Creek. A steep trail continues down to the tiniest of beaches at the creek mouth. Our trail crosses the creek on a wooden footbridge and passes through a 200-foot long tunnel cut into the cliffs.

At the cove are remains of a dock. The not-so-placid waters of the cove stir the seaweed about as if in a soup and you wonder how boats actually moored here. If you follow the crumbling cliffside trail from the dock to the end of the point, you may glimpse a sea otter.

Return the same way.

Cone Peak Lookout Trail (M-6)

Coast Ridge Road to Cone Peak; 4 miles RT; 1100' gain

Cone Peak, a geographical landmark to coast travelers for over a hundred years, is the most abrupt pitch of country along the Pacific Coast. It rises to 5,155 feet in about 3½ miles from sea level. On a clear day in winter, as you stand on Sand Dollar Beach, the snow-covered peak is a stirring sight.

Botanically, Cone Peak is a very important mountain. On its steep slopes botanists Thomas Coulter and David Douglas discovered the Santa Lucia fir, considered the rarest and most unique fir in North America. (Tree lovers know that when names were attached to western cone-bearing trees, Coulter's went to a pine, Douglas' to a fir.) The spire-like Santa Lucia fir, or Bristlecone fir, is found only in scattered stands in northern San Luis Obispo and southern Monterey Counties in the Santa Lucia Mountains. Typically, this fir occurs above the highest Coast redwoods (about 2,000 feet) within mixed evergreen forest. Santa Lucia fir concentrates in steep, rocky, fire-resistant spots at elevations from 2,000 to 5,000 feet.

Directions to trailhead: From Highway 1, 4 miles south of Lucia and just south of Kirk Creek Campground, 8¾ miles north of Gorda, turn east on Nacimiento-Fergusson Road. This road provides dramatic coastal views as it ascends sharply 7 miles to Nacimiento Summit. At the signed junction at the summit, turn left on graded Coast Ridge Road (Cone Peak Road) and follow it 5.2 miles north along the ridge to the signed trail junction on the west side of the road. Parking is adequate for a few cars. (Warning: During the rainy season Cone Peak Road may be closed for periods.)

The Hike: The well-graded trail ascends through oak woods. Soon the trail begins a series of steep switchbacks through brush. You'll enjoy views of Santa Lucia fir and Coulter pine. As the trail gains elevation, sugar pine, with its characteristic long cones predominates.

Hikers reach a signed junction 1¾ miles from the trailhead. (A trail leads west 1¼ miles down to steep, deeply-shaded Trail Springs Camp (3800'). For an interesting loop around Cone Peak, you can pick up the Gamboa Trail and ascend another 1¼ miles to the Coast Ridge Trail. Hikers then follow Coast Ridge Trail to its junction with Coast Ridge Road and follow the road a mile back to your car.)

From this junction, the main trail ascends a final ¼ mile eastward to the fire lookout atop Cone Peak summit. The lookout is staffed during fire season. Views of the valleys to the east and coastline to the west are yours. Spreading before you is a panorama of peaks: Pinyon Peak, Ventana Double Cone, Junipero Serra Peak, Uncle Sam Mountain.

Return the same way or hike the optional loop through Trail Springs Camp.

Prewitt Loop Trail (M-7)

Departing and arriving from Pacific Valley Forest Service Station
13 miles RT; 1800′ gain

Emil White's out-of-print classic, *Big Sur Guide,* has words of praise for most landmarks along Highway 1, until arriving at Prewitt Creek: "There are more than 30 bridges on this highway and the true historian may have something to say about each of them. We, however, have nothing to say about Prewitt Creek Bridge."

Prewitt Loop Trail offers a fine day hike, with good views of Big Sur's south coast. Added incentives to hike this trail include the fact that it's one of the few loop trails in the National Forest, and one of the few trails to receive anything like regularly scheduled maintenance.

Directions to trailhead: The trail starts behind the Pacific Valley Forest Service Station, just south of Prewitt Creek on the east side of Highway 1. Those driving from the north will find it a short distance north of Sand Dollar Picnic Area. Parking is by the fire station.

The Hike: Walk up the driveway behind the fire station to the signed Prewitt Loop Trail, which climbs above the fire station and a water tank. The trail joins a fire road for a few moments before resuming on the foot trail. Looking down on Pacific Valley, you'll notice there's nothing very valley-like about it. Almost any blanket-sized flat spot in Big Sur is signified with the word "valley."

You ascend on well-graded trail along Prewitt Creek watershed, changing coastal ecosystems with elevation: first grassland, then scrub, then woods, until 3 miles from the trailhead, you come to a spring. This spring, which feeds Prewitt Creek, is dependable for three seasons, chancy in summer.

Continuing the ascent, hikers are treated to fine views of the eastern ridges and down into Prewitt Canyon. The trail climbs along Plaskett Ridge, named for turn-of-the-century mailman Ed Plaskett who, with his departmental mule Jim, used to carry mail from the coast to Jolon and back along the then-tortuous Nacimiento-Fergusson Road.

The trail reaches its high point, nearly 2,000 feet above sea level, then crosses Prewitt Creek and begins descending the north side of the creek's watershed. During the wet season, waterfalls cascade down the canyons. On the grassy north slopes of the canyon, springtime brings fields of lupine and California poppy. Sand Dollar Beach, the largest beach on the Big Sur coast, can be glimpsed to the west. 2½ miles after crossing Prewitt Creek, you'll arrive at Stag Camp, located on a hillside. 1 stove, water from a hose.

Continuing its moderate descent of the north canyon slopes, the trail passes through meadowland. Near trail's end, hikers cross a pasture and arrive at a gate by Highway 1. The trailhead at Pacific Valley Station is a ¼ mile walk south along the highway.

Mill Creek Trail (M-8)

Nacimiento Road to Mill Creek; 1 mile RT; 100′ gain

Nacimiento Road to Nacimiento Summit; 12 miles RT; 2000′ gain

Mill Creek was the site of a turn-of-the-century logging operation. A mill located on the creek shaped the redwood logs. The sawn wood was then shipped down a rough wagon road to the creek mouth. At Mill Creek Landing, crane, cable and windlass lowered the lumber from the precipitous cliffs to waiting ships.

The usually overlooked and somewhat obscure Mill Creek Trail provides a delightful, but rugged day hike. Stray posts, shakes, and beams, along with bits of machinery and other remains of the logging operation are strewn about the forest floor. The first third of the trail is passable, the second third requires some creekside improvisation, the final third is easy-to-follow fire road. The trail, where it's not washed out, requires many creek crossings. For the most part, it remains in the bottom of redwood-shaded Mill Creek Canyon.

No doubt those hikers wishing to take it easy, will descend Mill Creek Trail from Nacimiento Summit and catch a ride back up the mountain.

Directions to trailhead: From Highway 1 just south of Kirk Creek Campground, turn east on Nacimiento-Fergusson Road, following it for a little over a mile to the signed Mill Creek Trail on your right. On the left, opposite the trailhead at a hairpin turn in the road, is a turnout for parking.

The Hike: Hike up a grassy slope, glancing over your shoulder at the Pacific panorama behind you. Pass through a barbed wire cattle gate and follow the well-graded trail along chaparral slopes. Soon you'll descend into the cool, moist, fern-covered canyon. Surely more banana slugs crawl along this trail than along any other California coastal trail. In a short time you'll reach Mill Creek.

The trail soon grows faint; however, traces of the old logging road are visible. For approximately the next two miles, watch for the blue Forest Service flags tied to trees and bushes; they'll help you stay on the nearly extinct road and locate the numerous stream crossings. The flags cease at the junction of a side canyon on the left with Mill Creek Canyon. A creek cascades down this steep and beautiful side canyon and joins Mill Creek.

For approximately the next 1½ miles it's rough going; pick your own route along the creek. Occasional stretches of washed-out trail help a bit. Connector trails, branching off from Mill Creek, shown on Forest Service maps, are so obscure you'll no doubt fail to notice them. After passing the remains of the old sawmill, it's another ¼ mile up-creek to a final creek crossing and the intersection with the jeep road leading to Nacimiento Summit.

At this final creek crossing, what looks like the main Mill Creek Canyon continues east, but you'll follow the jeep road as it climbs out of the canyon in a northerly direction. It's a stiff climb, but the jeep road is well graded. Soon you'll pass the crumbling remains of an old schoolhouse, which served the children of the loggers. The jeep road joins another canyon of the Mill Creek watershed, and follows it through mixed forest. You'll reach the Coast Ridge Road at its intersection with Nacimiento-Fergusson Road at Nacimiento Summit.

More coastal hiking in...

SAN DIEGO COUNTY

Coastal wetlands, including salt marshes, shallow water lagoons, and tidal mudflats, are vital habitats for certain fish, birds, and crustaceans. Many migratory birds stopover at wetlands, which also serve as vital habitat for two native Californian birds, the California clapper rail and the California least tern.

San Diego County offers a half dozen lagoons to explore. Most are bisected by the San Diego Freeway and are but remnants of far more extensive wetlands. Hikers may improvise a route up beach along the California Coastal Trail to reach these areas or simply select a lagoon and hike away. Bring your binoculars and favorite field guide.

La Salina Lagoon, a paved path leads along the banks of this tiny wetland. Shorebirds can be viewed from the path.

Buena Vista Lagoon hosts 200 varieties of waterfowl. Several viewpoints are on the perimeter of this state reserve.

Aqua Hedionda Lagoon, a walkway leads to a sandy beach at this recreation-oriented lagoon. Swimming, water skiing, fishing.

Batiquitos Lagoon, opposite South Carlsbad State Beach, is one of several unprotected areas serving as a refuge for endangered aquatic birds.

Penasquitos Lagoon, a feeding and resting place for native and migratory birds. Good views of lagoon and encroaching civilization are yours from trails in Torrey Pines State Reserve.

Border Field Federal Wildlife Refuge, located at the mouth of the Tijuana River; much wetland wildlife can be observed in this estuary.

Torrey Pines State Reserve, which protects the rarest of California pines, offers a few short, but pretty trails. Among them:

Beach Trail runs down the sandstone cliffs to a tide pool area.

Guy Fleming Loop Trail visits the Torrey pines.

Parry Grove Trail is a nature trail interpreting the coastal sage scrub ecosystem.

ORANGE COUNTY

The Santa Ana Mountains stretch the entire length of Orange County's perimeter, roughly paralleling the coast. This coastal range, much of which is protected by the Cleveland National Forest, is only about twenty miles inland and the western slopes are often blanketed with fog. The Santa Anas can be a wonderful region to hike, particularly in the cooler months—October through May. It's possible Santa Ana Trails could be connected to or become part of the California Coastal Trails.

You can obtain a Forest Service map from the Trabuco District Office located at Room 526, New Federal Building, 34 Civic Center Plaza, Santa Ana. You can also find trail information at San Juan Station on Ortega Highway.

Two nice trails to check out are:

Chiquito Basin Trail takes you past a sparkling little fall, over brushy hillsides and oak-studded slopes to shady Lion Canyon Creek.

Holy Jim Trail, runs along Holy Jim Creek to Holy Jim Waterfall and climbs to Bear Springs and Santiago Peak.

LOS ANGELES COUNTY

The Santa Monica Mountains are the only relatively undeveloped mountain range in the U.S. that bisects a major metropolitan area. They are a wilderness within an hour's drive of six million people, and stretch all the way from Griffith Park in the heart of Los Angeles to Point Mugu, fifty miles away. The range is twelve miles wide at its broadest point and reaches an elevation of a little over 3,000 feet. Large stretches are open and natural, covered with chaparral and oak trees, bright in spring with tulips, dainty white chamise and scarlet columbine.

The new Santa Monica Recreation Area is not one large area, but a patchwork of state parks, county parks and private property still to be acquired. The network of trails through the Santa Monicas is a rich pastiche of nature walks, scenic overlooks, fire roads and horse trails leading through diverse ecosystems: native tall grass prairies, savannas, yucca covered hills, and springs surrounded by lush ferns. Hopefully, the Backbone Trail, running the length of the range from Will Rogers Park to Point Mugu State Park will soon be completed. The Backbone Trail would then replace a beach route as a component of the California Coastal Trail.

For more information contact the Santa Monica Mountain National Recreation Area Headquarters: (213) 888-3770.

Malibu Creek State Park at 28754 Mulholland Dr., Agoura, (213) 991-1827, has volcanic peaks, woodlands, a lake and waterfall. 15 miles of trail explore the park. The **Malibu Creek Trail** travels along Malibu Creek to Malibu Lake Dam.

Topanga State Park at 20825 Entrada Rd., Topanga, (213) 455-2465. 35 miles of park trails comb the chaparral and meadows. **Eagle Rock Loop Trail** offers panoramic views of the ocean and the San Fernando Valley from Eagle Rock.

Will Rogers State Historic Park, 14235 Sunset Blvd., Pacific Palisades, (213) 454-8212, has hiking and equestrian trails touring the former ranch of the cowboy humorist. **Inspiration Point Trail** offers fine coastal views.

* * *

The Santa Catalina Island Conservancy, a non-profit foundation, owns 86% of the 28 mile long, 8 mile wide island and manages it as a preserve. Permits are required for hiking and camping in the interior and are available in Avalon at the Conservancy office, 206 Metropole Avenue, (213) 510-1421 and at the Information Center, 302 Crescent Avenue, (213) 510-2500; at the Cove and Camp Agency at Two Harbors, (213) 510-0303. Permits for hiking are free; however, backpackers pay the same (high) fee as car campers and must camp in designated campgrounds, of which there are five. Reservations are required.

Transportation to Catalina:

Catalina Island Cruises departs daily from San Pedro to Avalon. Limited service to Two Harbors. (213) 832-4521. P.O. Box 1948, San Pedro 90733

Long Beach/Catalina Island Cruises runs daily from Long Beach to Avalon. (213) 775-6111, 330 Golden Shores Blvd., Long Beach 90802

For additional information and a trail map, contact:
Los Angeles County Dept. of Parks & Recreation, Catalina Island, P.O. Box 1133, Avalon 90704; (213) 510-0688

Day hikes in the Avalon area are possible by seeking out the half dozen or so jeep trails and fire roads. Two nice ones to get you started:

Hour Trail leads from Catalina Stables to the Botanic Garden.

Divide Road journeys to East Mountain (1480′) providing panoramic views of both sides of the island.

VENTURA COUNTY

If you crave isolated ocean-sprayed coastlines that are unspoiled by any modern comforts—and uninhabited by humans—the Channel Islands are a terrific hiking choice. San Miguel and Anacapa Islands have been described elsewhere in this guide. Santa Barbara and Santa Cruz Islands also offer limited hiking opportunities. Before sailing out to the volcanic masses, drop in at the Channel Islands National Park Headquarters, Channel Islands Harbor, 1901 Spinnaker Drive, Ventura, (805) 644-8157.

If you lack a boat, the best way to visit one of the Channel Islands is by joining an organized expedition. The Cabrillo Marine Museum in San Pedro, the Natural History Museum of Los Angeles County and the Santa Barbara Museum of Natural History sponsor occasional trips. Private companies include Island Packers Co., Box 993, Ventura 93002; (805) 642-1393 and Santa Barbara Island Cruises, c/o Sea Landing Sportfishing, Breakwater, Santa Barbara, 93109, (805) 963-3522. The Nature Conservancy takes trips to Santa Cruz Island; 735 State Street, Suite 201, Santa Barbara, (805) 962-9111.

Tiny and austere Santa Barbara Island, a 635-acre remnant of an ancient volcano, has a network of trails including the self-guiding **Canyon View Nature Trail**. **Saddle Trail** crosses the middle of the island and connects with three more trails: **Signal Peak Loop,** which travels past giant coreopsis thickets to views of a rocky shore where sea lions haul out; **Elephant Seal Cove Trail** with views of these chubby pinnipeds on the beach below; and **Arch Point Loop,** which goes out to the lighthouse.

Santa Cruz Island is the largest island in the National Park and the one most similar to the mainland. It's densely wooded and has peaks rising to over 2,400 feet. Hikers are discouraged from "wandering around," because the Nature Conservancy, which manages the island, is trying to rehabilitate and preserve it. A good way to see Santa Cruz is to take a Conservancy sponsored trip.

The Ojai District of the Los Padres National Forest offers some exciting trails in the Topa Topa Mountains and White Mountains, as well as along such waterways as Sespe Creek and Piru Creek. Two favorites of local hikers:

Piedras Blancas Trail featuring some fantastic white rock formations.

Sespe Creek Trail travels through a handsome gorge in the midst of the Condor Sanctuary.

For more hiking information, contact: Ojai District, Los Padres National Forest, 1190 Ojai Ave., Ojai 93023; (805) 646-4348

SANTA BARBARA COUNTY

Most of the Santa Barbara Backcountry, a land of great gorges and sandstone cliffs, is in the Los Padres National Forest. Together Santa Barbara and Ventura Counties have more than a million acres of forest land. The Santa Barbara District includes the Santa Ynez and San Rafael Mountain ranges.

The Santa Ynez Mountains, which just a few million years ago rose from the sea, are of particular interest to the coastal hiker. They extend almost fifty miles from Matilija Canyon on the east to Gaviota Canyon on the west. Trails mainly follow the canyons above Santa Barbara and Montecito. They start in lush canyon bottoms, zig-zag up the hot dry canyon walls, and follow rock ledges to the crest. From the top, hikers enjoy seeping views of the Pacific, Channel Islands, and coastal plain.

More information about Santa Ynez Mountain trails can be obtained from Los Padres National Forest Headquarters, 42 Aero Camino, Goleta, CA 93017, (805) 968-1578.

Besides the Seven Falls and Gaviota Peak Trails described in this guide, other attractive and typical Santa Barbara foothill trails include:

San Ysidro Trail, explores the many aspects of San Ysidro Canyon and provides fine coastal views.

Rattlesnake Canyon Trail travels past quiet pools in Rattlesnake Canyon up to Gibraltar Road, offering an unobstructed view of the South Coast, Santa Cruz and Anacapa Islands.

SAN LUIS OBISPO COUNTY

The southern Santa Lucia Mountains stretch across much of San Luis Obispo County. Parts of the range are protected by the Los Padres National Forest. Those parts under Forest Service management resemble a chain of green islands in a sea of private land on USFS maps. The green islands are separated from the huge Big Sur and Santa Barbara Backcountry sections of the Los Padres.

The Santa Lucia Wilderness, established in 1978, consists of 21,500 acres of steep slopes surrounding Lopez Canyon. An all-year stream flows through Lopez Canyon down to Lopez Lake. The trail through Lopez Canyon gives hikers an opportunity to view a diverse streamside habitat. The Wilderness Area is accessible from the **Lopez Trail, Big Falls Trail,** and **Little Falls Trail.**

Further information on the Santa Lucia Mountains can be obtained from: Santa Lucia District, Los Padres National Forest, 1616 Carlotti Drive, Santa Maria 93454; (805) 925-9538.

Also by John McKinney

DAY HIKER'S GUIDE TO SOUTHERN CALIFORNIA
52 trails within 100 miles of Los Angeles

$8.95 plus .75 shipping

(Calif. residents please add 6% sales tax)

At your local bookstore or order direct from:

CAPRA PRESS, Box 2068-A, Santa Barbara, CA 93120